# PHILOSOPHICAL INTRODUCTION TO PROBABILITY

CSLI Lecture Notes
Number 167

# PHILOSOPHICAL
## INTRODUCTION TO
# PROBABILITY

MARIA CARLA GALAVOTTI

CSLI
PUBLICATIONS

Copyright © 2005
CSLI Publications
Center for the Study of Language and Information
Leland Stanford Junior University
Printed in the United States
09 08 07 06 05     1 2 3 4 5

*Library of Congress Cataloging-in-Publication Data*

Galavotti, Maria Carla.
Philosophical introduction to probability / by Maria Carla Galavotti.

p.  cm. – (CSLI lecture notes ; no. 167)

Includes bibliographical references and index.
ISBN 1-57586-490-8 (pbk. : alk. paper)
ISBN 1-57586-489-4 (alk. paper)

1. Probabilities.  I. Title.  II. Series.

BC141.G35   2005
121′.63–dc22        2005000367
CIP

CSLI was founded in 1983 by researchers from Stanford University, SRI
International, and Xerox PARC to further the research and development of
integrated theories of language, information, and computation. CSLI headquarters
and CSLI Publications are located on the campus of Stanford University.

CSLI Publications reports new developments in the study of language,
information, and computation. Please visit our web site at
http://cslipublications.stanford.edu/
for comments on this and other titles, as well as for changes
and corrections by the author and publisher.

Cover art by Alberto Pratelli

*In memory of*
*Dick Jeffrey (1926-2002) and Wes Salmon (1925-2001)*
*unforgettable teachers and friends*

# Contents

***Opening remarks***      1

**1. *The notion of probability***

**1.1  A historical sketch**        7

The birth of probability        7
The dual character of probability        12
Jakob Bernoulli and direct probability        13
Nikolaus and Daniel Bernoulli        15
Thomas Bayes and inverse probability        17
Probability and social mathematics: Condorcet and Quetelet        18
The rise of contemporary statistics: Galton, Pearson, Fisher        21
The advent of probability in physics        25

**1.2  Probability and induction**        27

Francis Bacon        28
Induction as ampliative inference        29
Hume's problem of induction        31
Mill, Herschel, Whewell        34

**2. *The laws of probability***

**2.1  The fundamental properties of probability**        39

**2.2  Bayes' rule**        47

**2.3  Kolmogorov's axiomatization**        52

## 3. *The classical interpretation*
### 3.1 Laplace and the Principle of insufficient reason    57

Determinism    57
The 'Principle of insufficient reason'    60
The 'Rule of succession'    62
Expectation and certainty    64

### 3.2 Problems of the classical definition    66

## 4. *The frequency interpretation*
### 4.1 Robert Leslie Ellis    71
### 4.2 John Venn    74

Probability as limiting frequency    74
Criticism of the rule of succession    77
Probability and belief    78

### 4.3 Richard von Mises and the theory of 'collectives'    81

Von Mises' approach    81
Collectives    83
Randomness    85
Collective-based probability    87
Applications to science    89

### 4.4 Hans Reichenbach's probabilistic epistemology    91

Reichenbach's frequentism    91
The theory of posits    95
The justification of induction    98
Causality    99

### 4.5 Ernest Nagel's 'truth-frequency' theory    101

## 5. *The propensity interpretation*
### 5.1 Peirce, the forerunner    105
### 5.2 Popper's propensity interpretation    106

Falsificationism    106
The propensity interpretation of probability    109

A world of propensities        112

**5.3  After Popper**        114

Single-case and long-run propensity theories        114
Humphrey's paradox        118
Propensity as an ingredient of the description of chance
phenomena        121

**5.4  Digression on chance and randomness**        125

Historical remarks        125
Poincaré's views on chance        126
The riddle of randomness        128
Is chance objective?        132

**6. *The logical interpretation***

**6.1  Beginnings**        135

**6.2  The nineteenth century English Logicists:
De Morgan, Boole, Jevons**        136

Augustus De Morgan        136
George Boole        138
William Stanley Jevons        141

**6.3  John Maynard Keynes**        144

Probability as a logical relation        144
Rationality and the role of intuition        147
Analogy, relevance and weight        149
Ramsey's criticism        152

**6.4  William Ernest Johnson**        153

**6.5  Viennese logicism: Wittgenstein and Waismann**        158

Ludwig Wittgenstein        158
Friedrich Waismann        161

**6.6  Rudolf Carnap's inductive logic**        164

Two concepts of probability        164
The logic of confirmation        169
The turn of the Sixties        174

**6.7  Harold Jeffreys between logicism and subjectivism**  178

Bayesianism  178
The interpretation of probability  181
Probabilistic epistemology  184

**7. *The subjective interpretation***

**7.1  The beginnings**  189

William Donkin  189
Émile Borel  191

**7.2  Frank Plumpton Ramsey and the notion of coherence**  194

Degrees of belief and consistency  194
Ramsey, Keynes and Wittgenstein  200
Belief, frequency and 'probability in physics'  202

**7.3  Bruno de Finetti and exchangeability**  208

De Finetti's radical probabilism  208
Subjective Bayesianism  215
Criticism of other interpretations of probability  220
Indeterminism  223

**7.4  Some recent trends**  225

Richard Jeffrey's radical probabilism  225
Patrick Suppes' probabilistic empiricism  230

*Closing remarks*  235

*References*  239

*Index*  261

# *Opening remarks*

Everybody agrees that life is dominated by uncertainty. Being a tool that enables us to face uncertainty, probability is an essential ingredient of human knowledge, both in everyday life and science. It is therefore natural that probability should be of some concern to philosophy, and it is precisely from a philosophical standpoint that it is addressed in this book. Probability invests all branches of philosophical investigation, from epistemology to moral and political philosophy, and impinges upon major controversies, like that between determinism and indeterminism, or between free will and moral obligation, and problems such as: 'What degree of certainty can human knowledge attain?' 'What is the relationship between probability and certainty?' 'What is the meaning of chance and its place in science?'. It is therefore advisable to narrow down the subject of this book and be very specific on what it is about.

This book focuses on the foundations of probability, and more specifically on the central problem in that context, namely that of the interpretation of the notion of probability. This notion will be taken in the quantitative meaning today associated with it, that is as a quantitative notion expressible by means of a function that assigns a hypothesis a value ranging in the interval 0–1. The given hypothesis can be a prediction regarding the occurrence of a singular event, or the expression of a general law concerning the behaviour of a (finite or infinite) class of events.

Ever since it took shape in the mid-seventeenth century, probability has been the object of wide scale debate. However, of its two sides: the mathematical one and the philosophical one, the latter is far more controversial,

1

due to the fact that probability can be taken as referring to our beliefs concerning what is uncertain, as well as to fortuitous events themselves. But while probability in relation to uncertainty is a characteristic of human knowledge – whose intrinsic incompleteness and imperfection make it necessary to appeal to probability – in relation to fortuitous events probability represents an ingredient of the description of an indeterministic world. This twofold meaning of the term lies at the roots of the various interpretations of probability, which have occasioned an ongoing controversy.

While focusing on the problem of the interpretation of probability, the following pages concentrate on a number of authors who have made specific contributions to the clarification and development of probability, taken in the quantitative sense specified above. The reader should not be surprised to find that many renowned philosophers, such as John Locke and Immanuel Kant, who did not address the notion of probability as we conceive it, are not mentioned in this book. The same holds for the meaning and uses of probability before the modern concept was spelled out. Furthermore, little space is devoted to induction taken in a general, non-probabilistic, sense. An extensive treatment of this notion – which belongs in the history of western thought since its very beginning – would require another volume. Instead, the book focuses on the peculiar traits and epistemological implications of the various interpretations of probability, and an effort is made to highlight the differences between the perspectives embraced by the authors adhering to each of them. The picture that emerges is much more diversified than is usually thought, to witness to the richness of the foundational debate revolving around probability.

Chapter 1 has a historical character and is meant to introduce the reader to the evolution of modern probability from its birth in the seventeenth century onwards. This chapter includes a section on the relationships between probability and induction – two notions that are now seen as intertwined, but were long addressed separately.

Chapter 2 introduces the mathematical properties of the concept of probability in a simple fashion, that does not presuppose in the reader a mathematical background. The chapter includes an account of Bayes' rule and Kolmogorov's axiomatization, which represent remarkable steps in the evolution of probability.

Chapter 3 is devoted to the so-called 'classical' interpretation of probability, worked out by Pierre Simon de Laplace at the turn of the nineteenth century. Laplace's theory of probability became very influential, and dominated the literature on the topic for a long time. It also raised various problems that provoked a vast debate. Favoured by the progressive widening of the range of application of probability, other interpretations were put forth in the nineteenth century to cope with the difficulties besetting Laplace's theory.

While the notion of probability developed by Laplace focuses on the epistemic meaning of probability, which is regarded as a component of human knowledge, an alternative viewpoint stresses the empirical aspect of probability and defines it in terms of frequencies. This is the frequency interpretation of probability, which took shape during the nineteenth century, and in the twentieth century became most popular with scientists, especially physicists. Chapter 4 deals with this interpretation, concentrating on the work of Robert Leslie Ellis, John Venn, Richard von Mises, Hans Reichenbach and Ernest Nagel.

Frequentism faces a problem in connection with the interpretation of quantum mechanical probabilities. In an attempt to deal with this difficulty, Karl Popper advanced the so-called 'propensity' interpretation in the late fifties. This had been anticipated by Charles Sanders Peirce, but after Popper's work it gained an ample consensus among philosophers of science. Chapter 5 outlines Popper's propensity interpretation and the ensuing debate. This chapter includes a section on chance and randomness, in which Henri Poincaré's account of chance phenomena in terms of their complexity and instability is discussed, and attention is drawn on the relativity of the notion of randomness. Finally, the reader will find some remarks on the objectivity of chance and its implications for the determinism/indeterminism issue.

Chapter 6 is devoted to the logical interpretation of probability. In contrast with the frequentist outlook, logicists follow Laplace in regarding probability as an epistemic notion. However, in an attempt to proceed beyond Laplace, they borrow the conceptual apparatus of logic. The chapter outlines the conception of probability of a number of authors, including Augustus De Morgan, George Boole, Stanley Jevons, John Maynard Keynes, William Ernest Johnson, Ludwig Wittgenstein, Friedrich Waismann, Rudolf Carnap – whose theory of probability marks the climax of the logical interpretation – and Harold Jeffreys.

Chapter 7 deals with the subjective interpretation of probability, which is currently the most popular version of the epistemic approach to probability. While sharing with logicists the conviction that probability pertains to our knowledge, rather than to stochastic phenomena, subjectivists stray from the former in that they claim that a given body of evidence supports one and only one correct evaluation of probability, relative to a given hypothesis. According to subjectivists, probability evaluations reflect degrees of belief, whose determination depends on a number of factors including a variety of elements, in addition to the available information. Special attention is paid to the work of Frank Ramsey and Bruno de Finetti, the most outstanding representatives of this current in the last century. Some recent trends in Bayesian epistemology, as upheld by such authors as Richard Jeffrey and Patrick Suppes, will be discussed.

Given the introductory character of this book, an effort has been made to keep the presentation as simple as possible, to address a broad readership – possibly from different backgrounds – who share an interest in the foundational problems connected with the concept of probability. As I stressed at the outset, the implications of the subject matter of this book are so numerous that any claim to completeness must of necessity be abandoned. Consequently, non-beginner readers will find that a number of authors who contributed in some way or other to the debate on the foundations of probability have been barely mentioned or not mentioned at all, and the same holds for various aspects or implications of the problems addressed. And of course much more could have been said about the topics and authors discussed. Nevertheless I hope that, with all its limitations, the following book will be of some use to those concerned with the problems relating to the philosophical side of the notion of probability.

## Acknowledgments

This book originates from my volume in Italian *Probabilità* (Florence: La Nuova Italia Scientifica, 2000), of which it is an expanded and completely revised version. Encouragement to produce this book came from a number of colleagues and friends, whose appreciation for my work has proved an invaluable support. Among them, I wish to mention with all my gratitude Pat Suppes, from whom I learned much more than I could tell, and whose support for the project of this book has been its *conditio sine qua non*. I owe an equally deep debt of gratitude to the memory of Wes Salmon, who first stimulated my interest in probability long ago, when I was a graduate student at Indiana University, and to that of Dick Jeffrey, who a few years later introduced me to subjective probability: 'the real thing' as he called it in the title of his last book. Sure enough, with such teachers, this book should be much better than it is.

The final version of this book benefited from useful comments by Domenico Costantini, Roberto Festa, Paolo Garbolino, Donald Gillies,

Antonello La Vergata, Brian McGuinness and Nils-Eric Sahlin, whom I warmly thank. I wish to thank an anonymous referee for suggesting a way of improving the exposition of the laws of probability in Chapter 2. My sincere gratitude goes to Frederico De Oliveira Pinto, the landlord of 7 St. Eligius Street, Cambridge, where I resided for extended periods while writing this book, finding there the peaceful atmosphere I needed, and to Raffaella Campaner for editorial work on the manuscript. Warmest thanks to Alberto Pratelli for the drawing that appears on the cover.

# 1

## *The notion of probability*

### 1.1 A historical sketch

The birth of probability

Today probability is understood as a quantitative notion, expressed by a real number ranging in the interval between 0 and 1. Probability values are determined in relation to a given body of evidence, conveying information that is relevant to the facts whose probability is to be estimated. The notion of probability so conceived is reputed to have come into existence in the decade around 1660, in connection to the work of Blaise Pascal and Pierre Fermat.

Early on, probability was mainly taken to refer to the approval of an opinion on the part of experts. In an authoritative study focused on Thomas Aquinas, Edmund Byrne shows how in the Middle Ages probability was not so much linked to empirical evidence, as it concerned approval by people widely recognized as upright and respectable[1]. In the context of medieval thought probability, as opposed to truly scientific knowledge – namely the kind of knowledge amenable to demonstration – 'assumes its basic meaning as indicative of the relationship between an opinion and an argument (*probatio*) or arguments brought forth in its favour (hence, "provable" or perhaps "approvable")' (Byrne 1968, p. xxv). Byrne detects in Aquinas' work the

---

[1] On the basis of an accurate analysis of the uses of probability in the fifteenth, sixteenth and seventeenth centuries, Kantola (1994) argues against Byrne's tenet, and claims that in the late medieval moral thought the term 'probability' occurs with a meaning close to the modern one. Kantola's work was pointed out to me by Nils-Eric Sahlin.

coexistence of two different usages of the term 'probability', which is taken sometimes to refer to propositions, and sometimes to contingent facts, more precisely to the frequency with which the latter occur.

As pointed out by Ian Hacking in *The Emergence of Probability* (1975), this duality of meaning is peculiar to the notion of probability, and persisted in the passage from the 'prehistory' to the 'history' of this notion, which passage was characterized by the fact that probability ceased to be associated with the opinion of experts, and started to represent the degree of certainty to be ascribed to an event whose outcome is uncertain on the basis of the available information, taken to include evidence based on facts.

The birth of probability is often associated with an anecdote, reported by a number of sources as authoritative as Gottfried Wilhelm Leibniz and Siméon Denis Poisson. The latter claimed that '*A problem about games of chance proposed to an austere Jansenist by a man of the world was the origin of the calculus of probabilities*' (quoted from Hacking 1975, p. 57). The gambler in this story is the French gentleman Chevalier de Méré (1607-1684), a conspicuous figure at the court of Louis XIV, while the 'austere Jansenist' is the prominent scientist and philosopher Blaise Pascal (1623-1662), who was converted to the Jansenist doctrine and in the last years of his life devoted great effort to defend it against disbelievers. The problems posed by Méré were of the following sort: 'in throwing two dice, how many tosses are needed to have at least an even chance of getting double-six?' (*ibid.*, p. 59), and 'how should one divide the stakes among the players, if a game is interrupted after a certain number of trials?'. To solve the problems, Pascal involved the leading mathematician Pierre Fermat (1601-1665), and in 1654 the two exchanged a correspondence, which is considered to be the cradle of modern theory of probability[2]. In that same year, Pascal himself worked out his ideas on the matter in *Traité du triangle arithmétique*, published posthumously in 1665.

To be sure, problems of the kind were in no way new, as they had been analysed and given solutions early on. What is new to the treatment of Pascal and Fermat is their systematic approach, which paved the way to the search for

---

[2] What is left of this correspondence is to be found in Fermat (1679), pp. 179-88. An English version of it is to be found in David (1962), pp. 229-53. Three letters by Pascal to Fermat are published in Pascal (1963), pp. 43-8. On Pascal's work, see the *Revue de Métaphysique et de Morale*, vol XXX (1923), no. 2. A recent, competent account of Pascal's work, including a detailed exposition of his contribution to probability and the exchange between Pascal and Fermat, is to be found in Shea (2003). See also Hammond, ed. (2003) for an accessible guide to Pascal's work.

universal rules detached from particular problems, and valid for all chance phenomena. As a matter of fact, before Pascal and Fermat the modern notion of probability cannot be said to be totally absent. Various authors, including Luca Pacioli (1445-1514), Gerolamo Cardano (1501-1576), Niccolò Tartaglia (1499-1557) and Galileo Galilei (1564-1642) had already studied a clutch of similar problems, coming up with insightful solutions. Moreover, there is a widespread tendency to hold that in the Renaissance period the so-called 'low sciences', including medicine, alchemy, astrology and the earth sciences, contributed to the emergence of statistics and probability[3].

Going back to Pascal, it is noteworthy that he was so deeply convinced of the overarching power of probability, that he did not hesitate to apply it to God's existence. The argument known as 'Pascal's wager' – contained in the famous *Pensées* (published in 1670) – is precisely meant to support belief in God by means of considerations grounded on probability and utility. Pascal's wager can be summarized as follows: betting on God's existence – and therefore acting as if God existed – is reasonable, because it maximizes expected good. For, if we, totally undecided on whether to evaluate God's existence as more or less probable than God's non-existence, assigned to it a probability of one half, we should bet on it anyway, in view of the expectation of two lives in place of one. The bet would be convenient even if we assigned an infinitely small probability to God's existence, because this is counterbalanced by the expected value of an eternal life, which can be considered infinite. Pointing out the historical importance of this argument, Hacking maintains that the relevant passages of Pascal's *Pensées*

> had an important byproduct: they showed how aleatory arithmetic could be part of a general 'art of conjecturing'. They made it possible to understand that the structure of reasoning about games of chance can be transferred to inference that is not founded on any chance set-up (Hacking 1975, p. 63).

Moreover, he observes that 'the popularity of the *Pensées* made it a familiar fact that games of chance could serve as models for other problems about form of decision under uncertainty' (*ibid.*). Pascal's wager testifies to the fact that

---

[3] More on the history of probability before Pascal is to be found in Franklin (2001); in Chapter 1, 'Prehistory of probability theory', of Maistrov (1974); David (1955) and (1964). Hacking's reconstruction of the birth of the modern notion of probability is challenged, among others, by van Brakel (1976), Garber and Zabell (1979) and Schneider (1980).

considerations of utility and bets had been an essential component of the notion of probability since the beginnings of its history.

More evidence in support of this claims is offered by the work of another pioneer of modern probability. One year after Pascal and Fermat corresponded, the Dutch scientist Christiaan Huygens (1629-1695), during a journey to Paris, met Chevalier de Méré at the house of the Duke of Roannez, a good friend of Pascal, and came to know the work of Pascal and Fermat. Back in Holland, he started working on probability problems, and two years later published the first treatise on probability, *De ratiociniis in aleae ludo* (1657)[4]. The distinctive feature of Huygens' work is the extensive use of what is today called 'mathematical expectation'. This is based on the idea that the value of a game depends both on the probability of obtaining certain outcomes and the gain associated with each of them.

Given the importance of this concept, it seems worthwhile illustrating it through an example. In contemporary jargon, we say that mathematical expectation is the 'fair price' of a chance game, obtained as the sum of all products of the probability and the gain or loss attached to each possible outcome. Consider the following game: a regular die is thrown, and for each throw a stake is fixed, valued at a sum in dollars equal to the number appearing on the upper side of the die after it is cast. Given that the die has six sides, all equally probable, the mathematical expectation of this game amounts to

$$(1/6 \times 1) + (1/6 \times 2) + (1/6 \times 3) + (1/6 \times 4) + (1/6 \times 5) + (1/6 \times 6) = 21/6 = 3.5 \text{ dollars.}$$

This sum represents the fair price of the game, in the sense that a player who paid this much to take part in the game, in a long series of throws can expect to gain enough money to cover the initial expense. In other words, the mathematical expectation of a game, or the average gain connected to it, should correspond to the fair price to be paid in order to take part in the game.

Strictly intertwined with the preceding is the idea that betting can be adopted as a possible tool for the evaluation of probability. Accordingly, one's degree of belief in the occurrence of an event can be expressed by means of the odds at which one would be ready to bet. For instance, a degree of belief of 1/6 in the fact that an unbiased die will turn up 3, can be expressed by the will to bet at odds 1 : 5 – namely, pay 1 if the die does not turn up 3, and gain 5 if it does. The general idea is to value the probability of an event as equal to the price to be paid by a player to obtain a unitary gain in case the event occurs.

---

[4] In Huygens (1888-1950).

This whole bunch of concepts was later put at the foundation of what is today called the 'subjective' or 'personalist' interpretation of probability, to be dealt with in Chapter 7. Granted that a long tradition of studies on utility and decision theory has added much sophistication to the matter, Huygens can be considered a pioneer in this field, as no doubt he advanced ideas endowed with 'a singularly modern flavour' (Hacking 1975, p. 95)[5].

In the second half of the seventeenth century, the study of probability progressed enormously, also thanks to the combinatorial calculus, and progressively widened its scope of application. Great impulse to such development came from the application of the notion of *arithmetic mean* to such data as natality and mortality tables, registered for annuities and insurance policies. An important contribution to advancement in this field came from the publication in 1662 of the *Natural and Political Observations Made upon the Bills of Mortality* by the London businessman John Graunt (1620-1674), and of another essay on the same topic by the Dutch mathematician and politician Jan de Witt (1625-1672). A valuable contribution also came from Huygens, who devised some measures of life expectancies[6]. The extensive use of statistical measures soon pervaded other fields, like medical practice, legal decisions, and, last but not least, the physical and biological sciences.

Gottfried Wilhelm Leibniz (1646-1716) worked intensively on the application of combinatorics to legal problems, laying great stress on the importance of probabilistic reasoning. As pointed out by Lorraine Daston in her *Classical Probability in the Enlightenment*

Leibniz combatted Locke's nominalism [...] by defending the reliability of probabilistic reasoning. Again and again, he supported this claim with legal examples of sound, albeit nondemonstrative, reasoning (Daston 1988, p. 45).

Leibniz's trust in probability is testified by § 372 of the *Nouveaux essais sur l'entendement humain*, where he writes: 'I maintain that *the study of the degrees of probability* would be very valuable and is still lacking, and this is a serious shortcoming in our treatises on logic' (Leibniz 1704, 1996, § 372). As we will see in Chapter 6, Leibniz is considered a forerunner of the logical interpretation of probability. Quite apart from that, there is no doubt that he greatly influenced his contemporaries, including some members of the family

---

[5] See David (1962), Chapter 11 for more on Huygens.
[6] See Chapter 3 of Daston (1988) for more details.

of scientists and mathematicians Bernoulli[7], to whom we will soon return.

## The dual character of probability

Before recollecting the contribution of other prominent figures in the history of probability, it is worth calling attention to the peculiar duality characterizing probability. In Hacking's words, probability is

> Janus-faced. On the one side it is statistical, concerning itself with stochastic laws of chance processes. On the other side it is epistemological, dedicated to assessing reasonable degrees of belief in propositions quite devoid of statistical background (Hacking 1975, p. 12).

Such a duality of meaning – which, as noted above, has marked probability since its prehistory – persists in the work of those who shaped its modern conception. Suffice it to think of Pascal, who readily transposed probability from the realm of chance games to belief in God. These two meanings have coexisted and often mingled in the work of probabilists for a long time, giving way to a terminological distinction between chance, that is probability taken in an objective sense, and referred to stochastic phenomena, and epistemic probability, or degree of credence.

This duality of probability lies at the root of the philosophical problem of its interpretation and of the various schools that engaged in the vast debate which will be surveyed in the next chapters. Such schools were animated by the conviction that one meaning of probability – the objective or the epistemic – should be privileged over the other and put at the core of the definition of probability. Around the middle of the nineteenth century this absolutist tendency became predominant, after a long period in which the 'doctrine of chance' and the 'art of conjecture' peacefully coexisted.

The interpretation of probability must be kept separate from its mathematical features. The mathematical properties of probability, in fact, hold quite independently of the interpretation attached to it. To be sure, an interpretation of probability is adequate only if it satisfies the mathematical properties that form the content of the so-called 'probability calculus'. While today this distinction is generally agreed upon, in the early days of probability things were not so clearly stated, and mathematical problems were intertwined

---

[7] On this point, and more generally on Leibniz's contribution to the history of probability, see Daston (1988) and Hacking (1975).

with the philosophical aspects of the notion of probability, taken sometimes as objective chance, and sometimes as degree of belief.

## Jakob Bernoulli and direct probability

A decisive contribution to the development of the notion of probability came from the work of various members of the Bernoulli family, of Dutch origin but who later settled in Basel. The first Bernoulli to be mentioned here is Jakob[8] (1654-1705, also called Jacques), mathematician and physicist, who, partly together with his younger brother Johann (1667-1748, also called Jean), made outstanding contributions to various fields, including algebra, differential calculus and the theory of continuous series. Jakob's major work in probability is *Ars conjectandi*, published posthumously in 1713 by his nephew Nikolaus. Jakob's name is tied to the so-called 'Bernoulli's theorem' or 'weak law of large numbers'. Briefly, it says that, if $p$ is the probability of obtaining a certain outcome in a repeatable experiment, and $m$ the number of successes obtained in $n$ repetitions of the same experiment, the probability that the value of $m/n$ falls within any chosen interval $p \pm \varepsilon$ increases for larger and larger values of $n$, and tends to 1 as $n$ tends to infinity[9]. With his work, Bernoulli starts the analysis of what is usually called *direct probability*, that is – in today's jargon – the probability to be assigned to a sample taken from a population whose 'law' (what we have called $p$) is known. This kind of probability, and with it Bernoulli's result, is of the utmost importance, as it is, among other things, at the basis of sampling theory, which constitutes a major branch of statistics and a fundamental component of the methodology of empirical sciences.

Bernoulli's result is based on the concept of *stochastic independence*, which for the first time receives an unambiguous definition. The theorem holds for binary processes, namely processes that admit of two outcomes – like 'heads' and 'tails', or 'presence' and 'absence' of a certain property. Bernoulli's work also sheds light on the relationship between probability and frequency, by keeping separate the probability and the frequency with which the events of the considered dichotomy can theoretically occur in any given number $n$ of experiments, and sets the probability distribution over possible frequencies: 0, 1, 2, ... , $n$, usually called *binomial distribution*.

---

[8] In this as in other cases the spelling reported in *Dictionary of Scientific Biography* (Gillispie, ed. 1972) has been adopted.

[9] For an accurate exposition of Bernoulli's theorem, and more generally for his contribution to the theory of probability, see Maistrov (1974).

A further aspect of Jakob Bernoulli's work which is worth mentioning amounts to his analysis of evidence. He made an insightful distinction between 'pure' evidence, taken to support a given hypothesis, and 'mixed' evidence, supporting both a given hypothesis and its negation. As Bernoulli puts it:

> some arguments are *pure*, others are *mixed*. *Pure* arguments prove a thing in some cases in such a way that they prove nothing positively in other cases. *Mixed* arguments prove a thing in some cases in such a way that they prove the contrary in the other cases. Example. Suppose someone in a milling crowd is stabbed with a sword and it is established by testimony of reliable witnesses looking on from a distance that the perpetrator of the crime had on a black cloak; suppose further that among the crowd Gracchus along with three others is found wearing a cloak of this color. The black cloak will be an argument that Gracchus committed the crime, but it will be a mixed argument, since in one case it proves his guilt but in three cases his innocence, depending on whether he or one of the other three is the perpetrator; for one of them could not be the perpetrator without Gracchus by that very fact being innocent. If, however, in a subsequent interrogation Gracchus turns pale, his pallor is a pure argument: it proves Gracchus' guilt if it results from a guilty conscience; but it does not, on the other hand, prove the innocence if it has another origin. For it is possible that Gracchus could turn pale for another reason and still be the murderer (quoted from Sahlin 1986, pp. 254-5).

This distinction is the source of much debate in the literature on evidentiary value[10].

Bernoulli's work on direct probability was carried on by other eminent probabilists. Among them was Abraham de Moivre (1667-1754), a French Huguenot who emigrated to England after the revocation of the Edict of Nantes in 1685, and author of *De mensura sortis* (1711), later revised as *The Doctrine of Chances* (1718). He proceeds one step further with respect to Bernoulli's theorem, by proving that when the number $n$ of experiments tends to infinity, the binomial distribution approaches the continuous distribution that was later called *normal*. The result obtained by de Moivre for the case of dichotomic events with equal probability was to be generalised by Pierre Simon de Laplace

---

[10] The problem is discussed in Sahlin (1986) and Rabinowicz and Sahlin (1998), where the reader will find references to relevant literature. I am indebted to Nils-Eric Sahlin for drawing my attention to this aspect of Bernoulli's work.

(1749-1827) for the case of events with different, but not too different, probabilities. Further developments in the analysis of direct probability are due to Siméon Denis Poisson (1781-1840), author of *Recherches sur la probabilité des jugements en matière criminelle et en matière civile* (1837). With respect to the theorem of de Moivre-Laplace, which is based on the assumption that the examined cases have constant, but not too different, probabilities, Poisson proceeds a step further. He relates Laplace's work on direct probability to Bernoulli's theorem, by extending Laplace's result to the case of dichotomous events with different probabilities, and demonstrates the so-called 'law of large numbers', which roughly says that the binomial distribution, as *n* tends to infinity, approaches the normal distribution, even in the case of different probabilities. He also sets the *poissonian* distribution as a limiting case of the binomial distribution, for the case in which one of the two probabilities is very small, and the other very large.

The study of direct probability received great impulse in the nineteenth and twentieth centuries, especially by the Russian probabilists[11]. One of them, Pafnuty Lvovich Chebyshev (1821-1894) founded the Saint Petersburg school, which included among others Andrej Andreevich Markov (1856-1922), Alexandr Mikhailovich Lyapunov (1857-1918), and Andrej Nikolaevich Kolmogorov (1903-1987), on whose work something will be said in the next chapter. Substantial progress was also made by other probabilists, like the Frenchman Émile Borel (1871-1956) and the Italian Francesco Cantelli (1875-1966).

## Nikolaus and Daniel Bernoulli

Let us go back to the Bernoulli family, for a brief recollection of the work of Nikolaus (1687-1759, also called Nicholas), nephew of Jakob and Johann (was the son of their brother Nikolaus), mentioned above as the editor of Jakob's *Ars conjectandi*. Nikolaus corresponded with the French probabilist Pierre Rémond de Montmort (1678-1719), author of the second volume on probability ever published, which appeared in 1708 with the title *Essai d'analyse sur les jeux de hasard*. A follower of Jakob Bernoulli's work, Montmort accused de Moivre of plagiarism, claiming he had been the first to reach some of the results later spelled out by de Moivre[12].

---

[11] See Maistrov (1974) on the contribution of the Russian school.

[12] On Montmort's work and his disappointment at de Moivre see especially David (1962), Chapter 14 and Todhunter (1865, 1965), Chapter 8.

Nikolaus Bernoulli formulated the so-called 'Saint Petersburg problem', which was to be the object of much debate. A simple exemplification of the problem is the following: let us imagine betting on heads in a coin-tossing game under the condition that if heads turns up on the first toss, a certain player pays 10 dollars to the other, if heads does not turn up, at the second toss the sum at stake becomes 20 dollars, to become 40 dollars at the third toss, and so on doubling the stake at every toss. Suppose the game has to be continued until the coin turns up heads, the problem arises of calculating its mathematical expectation, for it is easily seen that such expectation 'is infinite, for there is a finite, though vanishingly small probability that even a fair coin will produce an unbroken string of tails, and the payoff always grows apace' (Daston 1988, p. 69). The idea of a game having an infinite price obviously collides with common sense, whence the paradoxical character of this problem, sometimes called the 'Saint Petersburg paradox'. The problem has called for much attention[13], despite the fact that, as remarked by Nikolaus Bernoulli himself, 'no reasonable man would pay more than a very small amount, much less a very large or infinite sum, for the privilege of playing such a game' (*ibidem*).

Another Bernoulli worked on the solution to this problem: Daniel (1700-1782), son of Johann and author of a number of writings on various topics, including mathematics, mechanics, medicine and probability. The most important of his essays on probability was the *Specimen theoriae novae de mensura sortis* (1738). Daniel Bernoulli's analysis of the Saint Petersburg problem led him to embrace a new perspective on the notion of expectation. In his hands, the latter ceases to be a purely mathematical concept, to be transposed to an economic framework in which risk and gain do not have an abstract value, but are linked to utility, as determined by practical circumstances. 'Moral' expectation takes the place of 'mathematical' expectation. This attitude accounts for the problems related to the 'marginal utility' of money, a well known issue today, that Daniel Bernoulli saw very clearly. In Daston's words, the question amounts to the fact that 'the plight of a poor man holding a lottery ticket with 1/2 probability of winning 20,000 ducats' is 'in no way symmetric to that of a rich man in the same position: the poor man would be foolish not to sell his ticket for 9,000 ducats, although his mathematical expectation was 10,000; the rich man would be ill-advised not to buy it for the same amount' (Daston 1988, p. 71). Starting from such

---

[13] For a discussion of the Saint Petersburg problem and the debate on it, see Jorland (1987).

considerations, Daniel Bernoulli attempted a solution to the Saint Petersburg problem based on treating utility as a logarithmic function of monetary wealth. In this way, as the game goes on, smaller and smaller values of mathematical expectation obtain. If Bernoulli's solution to the problem has not been considered satisfactory by a number of authors, no doubt his perspective is very innovative, and his notion of moral expectation has a direct bearing on contemporary theory of decision, game theory and subjective probability[14].

Daniel Bernoulli also analysed the distribution of errors of observation, coming to the conclusion that small errors are more probable than large ones[15]. In so doing, he laid the foundations of one of the most fruitful branches of statistics, namely the theory of errors, which reached its peak with the work of the multifarious genius of the mathematician and scientist Carl Friedrich Gauss (1777-1855). Gauss set the distribution of errors of measurement, which turned out to correspond to the law independently found by de Moivre and Laplace through the analysis of other problems[16]. *Gaussian* or *normal* distribution had an enormous impact on statistics and scientific methodology at large. It was applied not only to the treatment of experimental errors, but to a wide range of phenomena. The theory of errors was the object of extensive analysis also by the English mathematician Thomas Simpson (1710-1761), who opened a most important field, being the first to study continuous distributions of probability[17].

## Thomas Bayes and inverse probability

During the second half of the eighteenth century, an important result was obtained by Thomas Bayes (1701-1761), a Nonconformist minister at Tunbridge Wells, near London, elected Fellow of the Royal Society in 1742[18]. The result in question is contained in the famous 'An Essay towards Solving a Problem in the Doctrine of Chances', which was found among Bayes' papers after his death by his friend Richard Price (1723-1791), author of the

---

[14] More on Daniel Bernoulli is to be found in David (1962), Chapter 13.

[15] On this issue see Daniel Bernoulli (1777), whose English translation also contains a comment by Leonhard Euler (1707-1783), another great mathematician and probabilist, whose work is not mentioned here for shortage of space.

[16] The bibliography on Gauss, who is one of the greatest mathematicians of all times, is huge, and so is the list of his publications. For an intellectual biography, see Dunnington (1955).

[17] See Maistrov (1974), pp. 82-7 on Simpson's contribution.

[18] Sketches of Bayes' life are to be found in Barnard (1958) and Holland (1962).

Northampton mortality tables and a writer on politics, morals and economics. Price read Bayes' paper to the Royal Society on December 23, 1763. In it Bayes spells out a method for assessing *inverse probability*, or the probability to be assigned to a hypothesis on the ground of available evidence. Whereas by direct probability one goes from the known probability of a population to the estimated frequency of its samples, by inverse probability one goes from known frequencies to estimated probabilities. Inverse probability is sometimes called 'probability of causes', because it enables us to estimate the probabilities of the causes underlying an observed event. The analysis of inverse probability, later carried on by Laplace, was to become a major field within probability theory[19].

The importance of Bayes' result – which will be described and illustrated by means of an example in the next chapter – lies with the fact that it points to a way of evaluating hypotheses in a probabilistic fashion. In other words, Bayes devised a canon of inductive reasoning, which transforms an initial probability – assigned to a family of statistical laws on the basis of background information – into a final probability – obtained by updating initial probability in the light of newly acquired information. In Bayes' times induction and probability were considered separately, and no connection between them was established. Bayes' method combines them, thereby anticipating modern statistical inference. According to the school of statistics called *Bayesian*, Bayes' method can be regarded as the core of all statistics. Though this tendency is no doubt prominent among contemporary statisticians, it is by no means the only one.

## Probability and social mathematics: Condorcet and Quetelet

In addition to the progress brought to the field by the Bernoulli, Bayes, and many others, probability considerably widened its range of application in the eighteenth century, particularly in the realm of the moral and political sciences. An outstanding representative of such a tendency was the French mathematician Marie-Jean-Antoine-Nicolas Caritat, Marquis de Condorcet (1743-1794). After taking part in the French Revolution and playing a role in drafting the Declaration of Rights of 1789, Condorcet ended his days by self-

---

[19] For a remarkable reconstruction of the history of the notion of inverse probability see Dale (1991).

poisoning during the Terror, to escape the guillotine of the Jacobins, who suspected him to be a Girondin[20]. Condorcet drew a distinction between 'absolute' probabilities, obtained through the calculus of probability, and the 'belief' that guides human decisions. In an attempt to bridge the two, he appealed to Bayes' rule of inverse probability, and came up with some personal results. The importance of Condorcet's contribution to the mathematical theory of probability was disputed by the historian of probability Isaac Todhunter, who charged it with 'obscurity and self-contradiction [...] without any parallel' (Todhunter 1865, p. 352)[21]. Be that as it may, it is worth pointing out that Condorcet valued probability calculus as an ideal tool at the service of reason, apt to free it from error and to ground human belief on a solid and objective basis.

Condorcet is often remembered as the opener of a new research field, the so-called *social mathematics*, which was meant to produce a statistical description of society instrumental for a new political economy. The whole idea of social mathematics is inspired by the conviction that the sciences of man can also be given an empirical foundation. Like the natural sciences, they can admit of a rigorous and exact methodology, which is provided by the calculus of probabilities. This holds true of the human mind as well, which for Condorcet is modelled according to the operations of mathematics. Condorcet's attitude is thus summarized by Lorraine Daston:

> Condorcet assumed that enlightened men who naturally reckoned by inverse probabilities would always agree with one another: right reason admitted only one answer to any question. By public instruction in the results, if not the mathematical theory, of the calculus of probabilities, the circle of reasonable consensus could be widened to include everyone (Daston 1988, p. 218).

Condorcet exerted an influence on his contemporaries, including Laplace and Poisson, who vigorously supported the application of probability to human affairs.

One can say that by the end of the eighteenth century probability was applied to all fields of inquiry and it was regarded as an indispensable tool for

---

[20] For an intellectual portrait, including the list of publications of Condorcet, see Granger (1956, 1989).

[21] A more sympathetic discussion of Condorcet's ideas on probability, conducted at a more informal level than Todhunter's, is to be found in Daston (1988).

investigating scientific as well as moral phenomena[22]. At the turn of the century, the study of probability reached its apex with the work of Laplace, to be dealt with in Chapter 3. Not only did he work out a vast systematization of mathematical probability, but he developed a perspicuous and influential philosophical account of it.

Before we turn to some of the major developments of probability in the nineteenth century, let us pause for a few remarks on the notion of arithmetical mean. Already known to the Pythagoreans and utilised in the study of probability since its very beginning, this notion gained increasing relevance in the course of the seventeenth and eighteenth centuries. By the end of the eighteenth century the notion of arithmetic mean was extensively used in all fields, from actuarial tables to the study of nature. Not only did it serve the purpose of summarizing data in both a synthetic and precise way, it also entered into the theory of errors, where it combined with the notion of 'deviation' between the data and the mean value, as well as in the calculation of mathematical expectation. In the nineteenth century, the impact of the notion of arithmetical mean grew so much that it was elevated to the role of heuristic principle. This happened with the work of Adolphe Quetelet (1796-1874), professor of mathematics in Ghent, afterwards curator of the Royal Observatory in Brussels and permanent secretary of the Brussels Academy, and an enthusiastic supporter of the fruitfulness of statistics for the study of man. His most famous essay *Sur l'homme et le développement de ses facultés, ou Essai de physique sociale* (1835) started a new era of statistics, animated as it was by the conviction that the application of statistical methods to the study of man would give it the rigour and precision of physics.

According to Quetelet, in order to discover the laws governing human society one should proceed exactly as in physics; that is by observing as many cases as possible, and then calculating mean values. Averages provide the skeleton of the laws that govern man and society. For Quetelet the advantages of quantitative analysis were obvious: not only did it allow for the comparability of phenomena, it also paved the way to the search for causes, which constituted a necessary step towards improving the environment and social conditions of life. Quetelet's well known notion of *average man*, which occupies a pivotal role in his perspective, works as a heuristic device, paving the way to the study of populations. His analysis, however, extended beyond

---

[22] On the progressive application of probability to the study of natural and social phenomena between the eighteenth and nineteenth centuries, see Hacking (1990).

averages, to the variation of the populations analysed. This is pointed out by Stephen Stigler, who writes that

> l'*homme moyen* was the anchor for a group, and took its meaning *from* the group. As he [Quetelet] emphasized repeatedly, the mean was not enough; the dispersion about the mean, whether it be expressed as limits or in some other way, was essential to the comparison (Stigler 1999, p. 63).

An important innovation due to Quetelet is the application of the normal distribution outside of the realm of error theory. Having collected large masses of data, he was able to apply them successfully to the chest circumference of a population of soldiers. Borrowing Stigler's words, one can say that Quetelet's work 'helped create a climate of awareness of distribution that was to lead to a truly major advance in statistical methods' (Stigler 1986, p. 219)[23].

## The rise of contemporary statistics: Galton, Pearson, Fisher

Francis Galton (1822-1911), cousin of Charles Darwin (1809-1882), Gold Medal of the Geographical Society for his explorations in southwestern Africa, was a man of unflagging energy and very wide interests, who made substantial contributions to a number of fields, ranging from meteorology to anthropology, psychology, and above all statistics[24]. Among other things, he was the first to identify the anticyclone, and invented the method of personal identification based on fingerprints. His major works are *Hereditary Genius: An Inquiry into its Laws and Consequences* (1869) and *Inquiries into Human Faculty and its Development* (1883). Galton was inspired by an unshakeable confidence in the fruitfulness of mathematization in all fields of inquiry. Following up Quetelet's idea that various features of man conform to the normal distribution, Galton gathered a large number of data on biological and psychological phenomena, in order to proceed to a systematic comparison between them. Galton's foremost contribution to statistical methodology lies with the fact that, by focusing on the variation of data around the mean values, he was able to devise various measures of regression and correlation. Previously considered essentially a

---

[23] On Quetelet see also Porter (1986).

[24] For Galton's biography see Gillham (2001), which contains also some information on Galton's disciples Karl Pearson and Raphael Weldon. An invaluable source of information on Galton is still Karl Pearson, ed. (1914-30).

nuisance, the variability of data thus became an object of study of its own, and its analysis was the source of tremendous advances in statistics.

Galton's interest in heredity and variation (physical as well as mental) in mankind was closely connected with his campaigning for *eugenics*. By this term, which he coined in 1883, he meant

> the science of improving stock, which [...] takes cognisance of all influences that tend in however remote a degree to give to the more suitable races or strains of blood a better chance of prevailing speedily over the less suitable than they otherwise would have had (Galton 1883, pp. 24-5).

It can be said that it was Galton's fear of the degeneration of the English race and the falling birth rate of the upper classes that led him to launch a programme for collecting extensive data on heritable characters and their distribution in the population. As a strong believer in the power of 'nature' as opposed to 'nurture', he advocated the use of statistical techniques to put the study of variation on a firm basis and as a preliminary step towards the establishment of the laws of inheritance. Statistical methods were indispensable, and probability was to be – as the title of his Herbert Spencer lecture for 1907 goes – 'the foundations of eugenics'[25]. It must be remembered that, as in Galton's case, interest in eugenics and concern for the biological destiny of the nation was the proximate cause for many biologists and mathematicians to apply statistical methods of investigation to the study of the distribution of characters in human populations. This will be particularly evident in Karl Pearson's and Ronald Fisher's case.

Galton's programme was pursued and developed in the first place by his most zealous disciple: Karl Pearson (1857-1936), Galton Professor of eugenics at University College, London, who was one of the main forgers of contemporary statistics. In 1901, Pearson founded, together with another disciple of Galton, Raphael Weldon (1860-1906), and Galton himself, the journal *Biometrika*, which became to official organ of the new discipline of 'biometry': a quantitative branch of biology bound to lead to substantial developments in statistical methodology. Among Person's numerous works are *The Grammar of Science* (1892), and the collection in two volumes *The Chances of Death and Other Studies in Evolution* (1897). The former book, which made the name of Pearson known to a wider audience of scientists,

---

[25] See Galton (1907).

intellectuals and philosophers, testifies to a positivist attitude towards science, according to which scientific knowledge amounts to a net of functional relations among phenomena, to be expressed in quantitative terms. In Pearson's hands the analysis of regression and correlation made enormous progress. But his contribution to statistics goes far beyond that, for he can be considered the father of the so-called *significance tests*, a methodology for assessing statistical hypotheses against experimental data. The basic idea underlying testing methods is that of evaluating to what extent the consequences of statistical hypotheses are supported by experimental data, with the purpose of rejecting the hypotheses which do not fit the data. The methodology based on tests of significance was to become a major branch of statistical methodology.

A believer in State-socialism, although not properly a Marxist, Pearson thought eugenics should be applied on both national and international scale – in the latter case by an international body of experts sponsored by the League of Nations. Eugenics was the real solution to mankind's problems, the way to diminish suffering and to base social justice on biological grounds. This belief matched the conviction that nations would benefit from a more intensive reproduction of their better specimens, and the whole of mankind would benefit from the reproductive success of the higher races, and the disappearance of the lower ones[26].

Around the same period, other important figures were at work at the development of modern statistics. One should mention Francis Ysidro Edgeworth (1845-1926), Drummond Professor of political economy at All Souls College, Oxford, who is said to have anticipated various ideas later developed by Karl Pearson and Ronald Fisher[27]; William Sealy Gosset (1876-1937), who wrote a number of essays under the name of 'Student', and devoted special attention to tests based on very small samples[28], and Ronald Aylmer Fisher (1890-1962), second Galton Professor of eugenics at University College, London. In 1943, Fisher moved to Cambridge where he started a flourishing school of statistics, and in 1959 emigrated to Australia, where he ended his life. He published a number of works, including *Statistical Methods for Research Workers* (1925), *The Genetical Theory of Natural Selection*

---

[26] For a biography of Karl Pearson written by his son, see Egon Pearson (1938). See also the more recent Porter (2004).

[27] On Edgeworth's impact on the history of statistics see Stigler (1986) and (1999), and Porter (1986).

[28] On Gosset's life and work, see Egon Pearson (1990).

(1930) and *The Design of Experiments* (1935)[29]. In 1919, Fisher joined the Rothamsted Experimental Station, where he was charged with reassessing a large mass of agricultural and meteorological data. His work in this connection put him in an ideal position to develop new ideas on experimental design. They included a novel outlook on significance testing, which Fisher combined with what is considered his major accomplishment, namely the analysis of variance and covariance. Fisher's name is also associated with a statistical methodology for comparing hypotheses on the basis of a given body of data, using the notion of 'likelihood', which represents a measure of the chance of obtaining certain results, given a certain hypothesis[30].

Fisher made an important contribution to the 'genetical theory of natural selection', which grew out of his concern for eugenics[31]. As summarised by Ernst Mayr:

> Fisher's most important conclusion was that much of continuous variation, at least in man, is due to environmental influences. [...] Fisher always thought in terms of large populations, and although he was fully aware of the existence of errors of sampling, he thought that, owing to the selective differential of competing genes and to recurrent mutation, such errors of sampling would be in the long run of little evolutionary consequence, as is indeed true for large populations (Mayr 1982, p. 555).

This passage illustrates an important feature of Fisher's theory, namely the fact that his approach is based on large populations, and his conclusions hold in the long run. This matches with the idea that probabilities are essentially referred to frequencies, obtained from the observation of large masses of data. As a matter of fact, Fisher regarded probability as closely connected with frequency, and always opposed the idea that probability is an epistemic notion, which was upheld, for instance, by his contemporary Harold Jeffreys, whom he engaged in a long controversy[32].

The methodology of tests was extended in such a way as to be applied to the comparison between two alternative hypotheses by Egon Sharpe Pearson

---

[29] A portrait of Fisher, including also a bibliography of his works, is to be found in Kendall (1963).

[30] On Fisher's work see Hacking (1965), and Gigerenzer *et al.* (1989). See also Edwards (1972), containing a detailed discussion of the notion of 'likelihood'.

[31] This is argued by Olby (1981).

[32] On the controversy between Fisher and Jeffreys see Howie (2002).

(1895-1980), son of Karl and professor of statistics at University College, London, working together with the Polish statistician Jerzy Neyman (1894-1981), who emigrated first to London and then to Berkeley[33]. As described by Neyman himself, the leading assumptions underpinning this new methodology are:

> (i) the existence of two kinds of errors possible to commit while testing a hypothesis, (ii) the notion that these two kinds of error may be of unequal practical importance, (iii) that a desirable method of testing hypotheses must ensure an acceptably low probability, say $\alpha$, of the more important error, and (iv) that, point (iii) being satisfied with an acceptable $\alpha$, the probability of the less important error be minimized (Neyman 1977, p. 103).

The more important error, or error of the first kind, is that which consists in rejecting a true hypothesis, while an error of the second kind consists in accepting a false hypothesis. Tests of hypotheses have become an important branch of statistics, and are extensively used in many disciplines, especially econometrics[34]. Like Fisher, Neyman and Pearson also insisted on the frequency-based character of probability.

## The advent of probability in physics

As we have seen, the study of man's social and biological dimensions gave a decisive impulse to the development of statistics. But probability is an essential ingredient of physics too. The process by which probability gradually entered physical science, not only in connection with errors of measurement, but more penetratingly as a component of physical theory, can be traced back to the work of the Scottish traveller and botanist Robert Brown (1773-1858). In 1827, while studying the pollen of plants, Brown observed that the particles of pollen suspended in fluid continuously move along irregular paths. He then extended his observations to other powdered substances and reached the conclusion that all microscopic particles suspended in fluid are animated by such a restless motion. By clearly seeing that *Brownian motion*, as it is still called, was to be regarded as a general property of microscopic particles of all substances,

---

[33] On Neyman's life and personality, see Reid (1982). A number of Neyman's and Pearson's joint papers are collected in Neyman and Pearson (1967).

[34] More detailed accounts of Neyman-Pearson theory are to be found in Hacking (1965) and Seidenfeld (1979), which contains also an extensive discussion of Fisher's work. Mayo (1996) argues for a Neyman-Pearson based philosophy of science.

Brown paved the way to the analysis of physical phenomena characterized by great complexity.

Rapid progress in this connection was made in the second half of the nineteenth century, when the kinetic theory of gases and thermodynamics were developed by the Scot James Clerk Maxwell (1831-1879), the Austrian Ludwig Boltzmann (1844-1906) and the American Josiah Willard Gibbs (1839-1903). In these fields, probability is made indispensable by the need to use mean values to describe phenomena that are characterized by such a complexity as to make it impossible to analyse them in detail. Statistical mechanics is concerned with the formulation of laws connecting the mean values of those physical magnitudes that enter in the description of molecular motions. At this stage, probability had become a building block of physical theory. Suffice it to say that the second law of thermodynamics was given a probabilistic interpretation by Boltzmann since entropy is defined in statistical terms.

Around the years 1905-1906 Marian von Smoluchowski (1872-1917) and Albert Einstein (1879-1955) brought to completion the study of Brownian motion, of which they gave a description in probabilistic terms. More or less in the same years, the analysis of radiation led Albert Einstein and other outstanding physicists, including Max Planck (1858-1947), Erwin Schrödinger (1887-1961), Louis de Broglie (1892-1987), Paul Dirac (1902-1984), Werner Heisenberg (1901-1976), Max Born (1882-1970), Niels Bohr (1885-1962) and many others to formulate quantum mechanics. Through quantum mechanics probability penetrated science even further, as it became an ingredient of the description of the basic components of matter[35]. The intrinsic and non-eliminable character of probability within quantum mechanics – at least according to the so-called Copenhagen interpretation – is reaffirmed by Werner Heisenberg, author of the *Principle of uncertainty*. As Heisenberg puts it, in quantum theory 'what one deduces from an observation is a probability function' (Heisenberg 1959, 1990, p. 38). Furthermore, 'quantum theory actually forces us to formulate these laws [the laws of the theory of radiation] precisely as statistical laws and to depart radically from determinism' (Heisenberg 1958, p. 39).

Parallel to the making of contemporary science, different interpretations of probability took shape. The study of gases, and more generally of molecular

---

[35] A detailed and competent account of the advent of probability in physics is to be found in von Plato (1994). See also Porter (1986), Chapter 7 and Maistrov (1974), Chapter 4, Section 3. Volume 2 of Krüger, Gigerenzer and Morgan, eds., (1987) contains a number of articles of interest.

aggregates, suggested the adoption of a frequency notion of probability, and the same holds, as we have seen, for population genetics. Thus the frequency interpretation of probability came to the fore as the natural complement of statistical mechanics and genetics. However, quantum mechanics – as suggested by Heisenberg's passage – aims at describing the behaviour of single particles, and this clashes with the frequency interpretation, which, as Chapter 4 will illustrate in some detail, defines probability with reference to mass phenomena. To solve this problem, a special version of frequentism, called 'propensity theory' would be proposed. We will examine frequentism in Chapter 4 and the propensity theory in Chapter 5.

Frequentism reflects the empirical meaning of probability as the theory of aleatory phenomena. However, it did not totally obscure the epistemic meaning of probability, which Laplace put at the core of a theory that became so influential as to be labelled the 'classical' interpretation. The epistemic interpretation not only survived, but gave rise to two distinct currents, namely the logical and the subjective viewpoints. These will be dealt with in Chapter 6 and Chapter 7, respectively.

## 1.2 Probability and induction

Although they are today commonly seen as intertwined, induction and probability are different notions, with a different history. According to Larry Laudan[36], the encounter between induction and probability, at the origin of the probabilistic conception of induction that now prevails, occurred only in the second half of the nineteenth century. Up to that time, induction was regarded as aiming at certainty, not probability. With the notable exception of Thomas Bayes and Richard Price, who clearly saw the advantages of combining induction with probability, the usual tendency was to consider induction as a means of expanding knowledge, by setting general relations among large masses of experimental observations. The long tradition that flourished over the centuries on the qualitative notion of induction cannot be recollected in a few pages. Rather than embarking on such an enterprise – which would in itself require another book – let us now consider some figures and topics that seem particularly relevant to the issue.

---

[36] See Laudan (1973).

## Francis Bacon

Talking of induction, it is mandatory to mention Francis Bacon (1561-1626). A pupil of the Elizabethan élite, Bacon was educated at Trinity College, Cambridge, and trained as a barrister at Gray's Inn, London. His complicated public and political life registered various ups and downs. After a brilliant career that saw him first Attorney General, then Lord Keeper and finally Lord Chancellor in 1618 under King James I who appointed him Baron Verulam and Viscount of St. Albans, in 1621 Bacon lost his office after being accused of bribery, was condemned and even imprisoned for a few days in the Tower of London. After this sudden fall, Bacon spent the last five years of his life dedicating himself completely to study and writing. Coherently with his unshakable faith in experimental method, he conducted all sorts of experiments, and is said to have died of a cold congestion, contracted during an experiment of stuffing a chicken with snow[37].

Bacon's theory of induction is expounded in his famous *Novum Organum* (1620). This work makes no mention of probability, which as we saw emerged as a quantitative notion some decades later. The gist of Bacon's perspective is that any investigation starts by gathering the largest possible quantity of data regarding the object under study. Once a great mass of data has been accumulated, one proceeds to sort and organize them, with the help of the so-called 'tables'. Three kinds of tables have to be compiled: the 'Table of existence and presence', designed to keep a record of all those situations in which a certain property – Bacon calls it 'nature' – of a given phenomenon is present; the 'Table of deviation, or absence in proximity', which serves the purpose of separating the phenomenon in question from other similar, but distinct, phenomena; and the 'Table of degrees', or 'Table of comparison', recording the greater or lesser degree in which the property is possessed by the phenomenon, according to varying circumstances.

The compilation of the tables is the first step of the inductive method, which represents a guide to the search of properties characterizing phenomena, as well as a mean to rule out wrong hypotheses. So goes Bacon's text:

Now I have grown used to calling the office and function of these three

---

[37] The bibliography on Bacon – one of the most outstanding figures of the history of western thought – is huge. For a biography of Bacon see Fuller (1981); for a relatively recent account of his philosophy see Urbach (1987). A critical edition of Bacon's works under the direction of Graham Rees is being published as *The Oxford Francis Bacon*. So far the classic edition of Bacon's works was Bacon (1857-74).

tables the *Submission of Instances to the Tribunal of the Intellect*, and when *Submission* is complete, the work of *Induction* itself must be set in motion. For we must find, on *Submission* of every last one of the instances, a nature which is such that it is always present or absent when the given nature is, and increases or diminishes with it, and is [...] a limitation of a more general nature (Bacon 1620, 2004, p. 253; Book II, Aphorism 15).

Thus, according to Bacon, induction is a method for inferring general statements from the observation of many singular cases.

It is worth noticing that Bacon stressed the importance of taking into account negative as well as positive instances, and warned against what he regarded as a natural tendency to neglect negative evidence. In particular, he recommended to fight against 'the peculiar and permanent error of being moved and excited more by affirmatives than negatives', and maintained that the intellect 'ought to pay heed in a proper and systematic way to both equally; indeed in the true setting up of every axiom, the power of the negative instance is actually greater' (*ibid.*, p. 85; Book I, Aphorism 46). As this passage shows, Bacon's elaborate methodology includes some germs of falsificationism, as well as a full blown inductivism[38].

## Induction as ampliative inference

The concept of induction as a method for inferring general conclusions from particular premises originated in antiquity and systematized by Bacon, though reflecting a major application of this method, can by no means be taken to cover all of its uses. Therefore, today it is widely agreed that such a conception cannot be taken as a definition of inductive inference itself, which can also lead from particular premises to particular conclusions, from general premises to general conclusions, and even from general premises to particular conclusions. Simple examples of all such inferences are given by Brian Skyrms. To exemplify inductive inferences leading from particular premises to a particular conclusion, he mentions the following[39]:

Premises:

1. Car *A* is a Hotmobile 66 and car *B* is a Hotmobile 66.
2. Car *A* has the super-zazz engine and car *B* has the super-zazz engine.

---

[38] Thanks to Nils-Eric Sahlin for drawing my attention to this passage.
[39] The following examples are taken from Skyrms (1966, first edition), pp. 14 ff.

3. Car $A$'s engine is in perfect condition and car $B$'s engine is in perfect condition.

4. Both cars have the same type of transmission and the same final drive ratio.

5. Car $A$'s top speed is over 150 miles per hour.

Conclusion:

Car $B$'s top speed is over 150 miles per hour.

Inferences of this kind, leading from particular to particular, are actually very important, as they include arguments by *analogy*, which are extensively used in science and everyday life.

One can also have inductive arguments from general premises to general conclusions, such as the following:

Premises:

1. All students in this class are highly intelligent.

2. All students in this class are strongly motivated to do well.

3. No student in this class has a heavy work load.

4. No student in this class has psychological difficulties that would interfere with his course work.

Conclusion:

All students in this class will do well.

As observed by Skyrms:

inductively strong arguments with general premises and general conclusions play an important role in advanced science. For example, Newton's laws of motion were confirmed because they accounted for both Galileo's laws of falling bodies and Kepler's laws of planetary motion (Skyrms 1966, p. 14).

Inductive arguments can also go from general premises to particular conclusions:

Premise:

All emeralds previously found have been green.

Conclusion:

The next emerald to be found will be green.

Again, this exemplifies an important and widely used family of inductive inferences. Typically, one reasons from general premises to particular conclusions when predicting the next event of a certain kind, of which a number of instances have been observed.

The preceding considerations lead to the conclusion that the distinctive feature of inductive arguments does not stem from the kind of statements – general or particular – they include as premises and/or conclusions. The distinctive feature of induction is rather to be identified with its *ampliative* nature. This means that the conclusion of an inductive argument broadens the informative content of its premises. There follows that the truth of the premises of an inductive argument does not guarantee the truth of the conclusion. With an alternative formulation, one can say that inductive arguments are *non-demonstrative*.

The opposite holds for deductive arguments, which are *demonstrative* and *non-ampliative*. In the case of valid deductive arguments, we have that the conclusion cannot be false if the premises are true, which is what is usually meant by the claim that deductive inferences are 'truth-preserving'[40]. To be sure, *validity* only applies to deductive inference, not to induction. Instead of qualifying as 'valid' or 'invalid' like deductive arguments, inductive arguments can receive a greater or lesser degree of strength from the evidence supporting them. One usually speaks of the *inductive support* given by the premises to the conclusion of an inductive inference. In order to have some bearing on the strength of an inductive argument, however, the premises ought to mention some piece of information that is *relevant* to the conclusion. Inductive support depends crucially on the amount of relevant information brought by the premises in favour of the conclusion.

## Hume's problem of induction

The ampliative character of induction is at the root of the problem of justification, posed with unpaired clarity by David Hume (1711-1776) in his *A Treatise on Human Nature* (1739) and *An Enquiry Concerning Human Understanding* (1748). The Scottish philosopher's argument moves from a distinction between 'relations of ideas' and 'matters of fact'. Relations of ideas 'are discoverable by the mere operation of thought, without dependence on what is any where existent in the universe', while matters of fact 'are not associated in the same manner; nor is our evidence of their truth, however

---

[40] The distinction between (deductive) demonstrative and (inductive) non-demonstrative inferences is illustrated in Salmon (1966).

great, of a like nature with the foregoing' (Hume 1748, 1999, p. 108; Section IV, Part I, § 1-2)[41]. Relations of ideas can be said to be certain by intuition or demonstration, in other words they are *a priori* true, and the contrary of a true relation of ideas leads to contradiction. Instead, it is always possible to conceive the contrary of any assertion about matters of fact. '*That the sun will not rise tomorrow* is no less intelligible a proposition, and implies no more contradiction, than *that it will rise*' (*ibidem*). This is so because matters of fact are independent from one another, they do not stand in any logical relationship akin to that holding between relations of ideas.

Matters of fact are connected by means of the relation of cause and effect. Indeed, this allows for inferences that 'go beyond the evidence of our memory and senses' (*ibid.*, p. 109; Section IV, Part I, § 4). But it is precisely this ampliative shift that is not warranted and raises the problem of justification. Cause and effect are distinct entities, whose relation, far from being *a priori*, is produced from experience. But experience, however great, will never guarantee that the effect will follow from the cause. All we can say is that we have observed a great many times a constant conjunction between a certain cause and a certain effect, still we can easily conceive of such a conjunction breaking down in the future. The problem of causality then coincides with the problem of induction, and this amounts to the impossibility of finding a logical relationship providing a justification of the ampliative step from the cause to the effect.

Actually, Hume makes no claim to the effect that causal, or inductive, arguments should not be put to use. His scepticism regards the possibility of finding a necessary connection between cause and effect, not the fruitfulness of relying upon their observed constant conjunction to infer the future from the past. The justification for this is provided by 'custom' or 'habit'. In Hume's words:

> Wherever the repetition of any particular act or operation produces a propensity to renew the same act or operation, without being impelled by any reasoning or process of the understanding; we always say, that this propensity is the effect of *custom*. By employing that words, we pretend not to have given the ultimate reason of such a propensity. We only point out a principle of human nature (*ibid.*, p. 121; Section V, Part I, § 5).

---

[41] See also Hume (1739), Book I, Part III.

Induction is thus grounded upon habit, and receives a justification of a pragmatical kind, as the best guide to action. Its value derives from that of custom, of which Hume says that it 'is the great guide of human life. It is that principle alone, which renders our experience useful to us' (*ibid.*, p. 122; Section V, Part I, § 6).

In Section VI of the *Enquiry*, 'Of Probability', Hume makes a few remarks on probability. Observed frequencies, he claims, are at the basis of beliefs regarding future events:

> it seems evident, that, when we transfer the past to the future, in order to determine the effect, which will result from any cause, we transfer all the different events, in the same proportion as they have appeared in the past, and conceive one to have existed a hundred times, for instance, another ten times, and another once (*ibid.*, p. 132-3; Section VI, § 4).

Belief is also strengthened by the number of experiments which have been made, which confer more or less 'weight'. Hume argues that a great number of experiments, all showing positive cases, supports a degree of belief so high, that it can be taken as a 'proof' of the future occurrence of a given event. In this way laws of nature are established.

On the same basis, because of the total lack of evidence in their favour, Hume argues against miracles. In Section X, 'Of Miracles', he maintains that miracles do not exist: 'as a uniform experience amounts to a proof, there is here a direct and full *proof*, from the nature of the fact, against the existence of any miracle' (*ibid.*, p. 173; Section X, § 12). While reaffirming the link between experimental evidence and degree of belief, Hume came close to seeing the connection between induction and probability, but in fact he did not. On the contrary, Hume entertained a naive conception of probability, and did not grasp the substantial difference between a proof and a probability which is either vanishingly small or tending to its uppermost value 1, which makes his argument against miracles not just weak, but mistaken. A reply to Hume's argument on miracles came from Richard Price, mentioned earlier in these pages as the one who reported Bayes' memoir to the Royal Society. Unlike Hume, Price made use of mathematical probability and of Bayes' rule, coming to the conclusion that there is no proof against miracles, but only a small probability[42]. But Price's voice, like Bayes', was to remain isolated

---

[42] See Daston (1988), pp. 326-9 for a detailed account of Price's argument. See also Earman (2000), which contains a devastating criticism of Hume's argument against miracles, together with a reconstruction of both Price's contribution and the eighteenth

for a long time.

## Mill, Herschel, Whewell

In the first half of the nineteenth century induction drew the attention of many authors, including John Stuart Mill, John Herschel and William Whewell, who have been seen by some authors as the beginners of philosophy of science in England[43]. The importance attached to induction by John Stuart Mill (1806-1873), one of the major philosophers in the British empiricist tradition, is reflected by the title of his *A System of Logic Ratiocinative and Inductive* (1843). The goal Mill set himself was to proceed one step further than Bacon and answer Hume's critique of induction. Central to his perspective is the claim that 'the validity of all the Inductive Methods depends on the assumption that every event, or the beginning of every phenomenon, must have some cause' (Mill 1843, 1973, 1996, vol. VII, p. 562; Book III, Chapter xxi, § 1). Like all general principles attained by induction from experience, the law of universal causation serves as a basis for that multiplicity of inferences from particular to particular, that are made in scientific investigation. Towards the establishment of laws of nature, a pivotal role is played by comparison among phenomena. This is done with the aid of Mill's four canons: the 'Method of agreement', the 'Method of difference', the 'Method of residues', and the 'Method of concomitant variations'[44]. As Mill says:

> The four methods which it has now been attempted to describe, are the only possible modes of experimental enquiry – of direct induction *a posteriori*, as distinguished from deduction: at least, I know not, nor am able to imagine, any others. [...] These, then, with such assistance as can be obtained from deduction, compose the available resources of the human mind for ascertaining the laws of the succession of phenomena (*ibid.*, p. 406; Book III, Chapter viii, § 7).

When dealing with probability – or, as he says, the 'calculation of chances' – in the first edition of *A System of Logic* Mill criticizes Laplace's definition of probability. He writes:

---

century debate on miracles.

[43] This opinion is taken by David Hull, who also conjectures that reading these authors is likely to have influenced Charles Darwin; see Hull (1973), p. 115.

[44] For a discussion of Mill's method from a contemporary perspective, see Skyrms (1966, 2000).

In the cast of die, the probability of ace is one-sixth; not, as Laplace would say, because there are six possible throws, of which ace is one, and because we do not know any reason why one should turn up rather than another; but because we do know that in a hundred, or a million of throws, ace will be thrown about one-sixth of that number, or once in six times (Mill 1843, 1974, 1996, vol. VIII, Appendix F, p. 1141).

This passage might be taken to suggest that Mill embraced a view of probability as frequency. But in the seventh edition of *A System of Logic*, published in 1872, Mill ends Chapter xviii of Book III: 'Of the calculation of chances', with a footnote where, referring to John Venn's *The Logic of Chance* (1866), described as 'one of the most thoughtful and philosophical treatises on any subject connected with Logic and Evidence', he writes: 'In several of Mr. Venn's opinions, however, I do not agree. What these are will be obvious to any reader of Mr. Venn's work who is also a reader of this' (Mill 1843, 1973, 1996, vol. VII, p. 547; Book III, Chapter xviii, § 6). To be sure, Venn's approach to probability and chance – to be described in some detail in Chapter 4 – is at odds with Mill's view of induction and his idea that chance is imputable to lack of knowledge of causes – on which more will be said on Chapter 5. In the first edition of *A System of Logic*, Mill is also very critical of Laplace's application of probability to witnessing, and more generally to the moral sciences.

John Herschel (1792-1871), one of the most influential men of science of his age, wrote a great deal on a wide range of topics, including astronomy, optics and meteorology. His *Preliminary Discourse on the Study of Natural Philosophy* (1830) is meant as an updated version of Bacon's *Novum Organum*, in the light of subsequent scientific accomplishments. Herschel calls attention to the fruitfulness of the practice of extending inductions beyond the range of observed phenomena. The formidable heuristic power of induction lies in the possibility of extending hypotheses, which have been confirmed by experiments of a certain kind, to new fields. Special emphasis is put on inductive inferences from one magnitude to another:

> In extending our inductions to cases not originally contemplated, there is one step which always strikes the mind with peculiar force, and with such a sensation of novelty and surprise, as often gives it a weight beyond its due philosophical value. It is the transition from the little to the great, and *vice versa*, but especially the former. [...] Yet the student who makes any progress in natural philosophy will encounter

numberless cases in which this transfer of ideas from the one extreme of magnitude to the other will be called for (Herschel 1830, 1996, pp. 172-3).

The same aspect is stressed by another outstanding figure in eighteenth century history and philosophy of science: William Whewell (1794-1866), author of a number of works, including *The Philosophy of the Inductive Sciences* (1840). Whewell embraced an original perspective, which can be regarded as 'resulting from the application of the commingled categories of Kantianism and classical British empiricism' (Butts 1973, p. 54). At the cornerstone of Whewell's perspective we find the notion of 'consilience of inductions', inspired by the idea that prediction of facts unforeseen at the time when a certain hypothesis was formulated is an essential ingredient for the growth of scientific knowledge. As Robert Butts puts it, 'A consilience of inductions takes place when a hypothesis introduced to cover one class of facts is later seen to explain another, different class of facts. [...] usually, it is two *laws* that become consilient by being derived from one more general theory' (*ibid.*, p. 61). The role assigned to consilience is indeed crucial, for Whewell claims that 'the consiliences of our inductions give rise to a constant convergence of our theory towards simplicity and unity' (Whewell 1840, 1996, p. 239). Last but not least, consilience fosters belief in the truth of a theory.

It should not pass unnoticed that Whewell has a deductive view of scientific method, and to him induction is justified on deductive grounds. In his words:

> Deduction is a necessary part of Induction. Deduction justifies by calculation what Induction had happily guessed ... Every step of Induction must be confirmed by rigorous deductive reasoning, followed into such a detail as the nature and complexity of the relation ... render requisite. If not so justified by the supposed discoverer, it is *not* Induction (quoted from Butts 1973, p. 66).

Whewell upholds a 'hypothetical method', according to which the hypotheses that guide scientific research are suggested by intuition, and are accepted if they can account for observed phenomena. While embracing a position of this kind, he clearly opposes Mill's inductivism.

In the works of these authors, and many others who have not been mentioned here, probability is assigned a minor role, being essentially relegated to a pre-scientific stage of investigation. As repeatedly pointed out, induction is not regarded as being connected with probability.

However, the encounter between probability and induction was to happen soon, and after it took place problems like the confirmation of hypotheses and the justification of induction had to be re-stated. Once induction is re-formulated in probabilistic terms, Hume's problem becomes that of justifying the particular methods adopted to make inductive inferences. As we will see, the problem mingles with that of the interpretation of probability, which is the object of the rest of this book.

# 2

## The laws of probability

### 2.1 The fundamental properties of probability

In order to have a thorough grasp of the notion of probability, it is necessary to understand its mathematical properties. These form the content of the so-called 'probability calculus', which is deemed to hold for all probability functions, thereby defining their range of admissibility. As already remarked, such properties hold irrespective of the interpretation of probability adopted: all interpretations of probability should satisfy them. The following presentation of the fundamental properties of probability will be as simple and informal as possible[1].

A first distinction has to be made between *primitive* probabilities, referring to simple – or 'atomic' – events, and *complex* probabilities, obtained by combining primitive ones. Assertions regarding the probability that 'tomorrow it will snow', or that 'the next toss of this die will give a 6' refer to atomic events, whereas assertions regarding the probability that 'it will snow tomorrow and the day after tomorrow', or that 'the die will turn up 5 or 6' refer to complex events. The probability of complex events can be calculated on the basis of those of simple events, according to ways dictated by the calculus of

[1] There are a number of introductory accounts of probability conducted at a simple level. Those addressing a philosophical audience include that of Brian Skyrms, in the second edition of *Choice and Chance* (2000), Chapter 6; and that of John Earman and Wesley Salmon in Part III, called 'Probability', of Chapter 2, on 'The Confirmation of Scientific Hypotheses', of the handbook *Philosophy of Science*, ed. by Merrilee H. Salmon and others, (1992). A rigorous but accessible introduction by two of the most prominent probabilists of the Russian school, is to be found in Gnedenko and Khinchin (1962).

probabilities. In contrast, the probability of simple events can be calculated in various ways, none of which descends from the probability calculus, which is only concerned with complex probabilities, and does not involve a univocal method for determining primitive, or *initial* probabilities. As we shall see in the following chapters, the determination of initial probabilities is a highly controversial issue, at the core of the debate on the interpretation of probability. The two problems are strictly intertwined, so that different interpretations of probability involve different methods for the calculation of initial probabilities.

Though the laws of probability could be introduced without assuming a specific rule to fix initial probabilities, it nevertheless seems useful to specify a method for their determination, as this will allow us to consider some examples. The method that will be adopted for that purpose is known as 'classical', because it is usually associated with the classical interpretation of probability of Pierre Simon de Laplace – whose work will be reviewed in the next chapter. According to this method, the probability of a simple event is given by

the ratio of the number of favourable cases to the number of equally possible cases.

This method presupposes that it is possible to spell out all of the possible cases with respect to the occurrence of an event, which, taken altogether, constitute the 'space of events' associated with the given one. The method at hand needs the further assumption that all such cases are *equally* possible. This is a crucial and much debated issue, to which we shall return several times, but it should be added that there are many events that do not raise problems in this connection, namely the realm of games of chance, which – as we saw – was the cradle of the mathematical notion of probability. In all those cases in which both the space of events associated with a certain happening and the possibilities which are favourable to it can be identified, the method that has been introduced finds a straightforward application, and has a strong intuitive grip.

Let us suppose that in evaluating whether an event will occur or not, we judge that all possible circumstances are favourable. In this case, we say that the event in question will happen with certainty. If, on the contrary, we judge that all circumstances are against it, we say that the event in question will not happen. When some of the circumstances look favourable and some unfavourable to the occurrence of an event, we calculate that its probability is equal to the ratio of the number of favourable to the total number of possible circumstances. For instance, if I buy one lottery ticket out of 200,000, the probability of my ticket winning is 1/200,000, whereas if I buy all tickets my probability of winning is 200,000/200,000 – in other words, I will win for sure. If I do not buy any ticket, my probability of winning is 0/200,000 – for sure I will *not* win.

The first property characterizing probability can now be stated by saying that a probability value is a real number in the interval between 0 and 1. Let us denote by $p(A)$ the probability of the occurrence of a certain event $A$ – for instance, the probability that we obtain a 3 as a result of a non-biased die being tossed. We then have that

$$0 \leq p(A) \leq 1 \qquad (1).$$

When speaking of probability assignments, it would be proper to speak of the probability of a proposition describing an event, or better its occurrence or non-occurrence. This should be kept in mind, though in what follows we will speak of the probability of an event, for short. Now, if $A$ stands for an event which is certain

$$p(A) = 1 \qquad (2).$$

The upper and lower values of probability correspond respectively to logical necessity, or to tautological propositions, and to logical impossibility, or to contradictions. However, it would be mistaken to think that such values do not represent genuine probability values. In fact, they do not express impossibility in a physical or practical sense. For instance, when the probability that the Chernobyl nuclear plant would explode was evaluated equal to 0, this was not meant to say that its explosion was impossible, as facts unfortunately have shown. In this as in other cases, probability 0 means that, in the light of the available information, one cannot assign a significant value to probability. In other words, the relevant information available with regard to a certain event occurring is sometimes too scant, to allow for an evaluation of probability other than 0, but this does not mean that the event in question is impossible. Probability is always relative to a given body of evidence.

If a standard coin is flipped, it can turn up in two ways: heads or tails. We therefore assign probability 1/2 to each of these possibilities. Similarly, if a die is tossed, we evaluate 1/6 the probability that it will turn up on each side. According to the definition adopted, the first thing we have to do when assigning a probability value to an event, is to list all of its possible outcomes, or the 'space of events' associated with it. The space of events associated with coin flipping includes two possibilities, while the space of events relative to the tossing of a die includes six possibilities. Let us now take the case in which a coin is flipped three times in succession. The space of events associated with this event – indicating heads by 'H' and tails by 'T' – is the following:

| | | | | |
|---|---|---|---|---|
| 1. HHH | $e_1$ | 5. THH | $e_5$ |
| 2. HHT | $e_2$ | 6. THT | $e_6$ |
| 3. HTH | $e_3$ | 7. TTH | $e_7$ |
| 4. HTT | $e_4$ | 8. TTT | $e_8$ |

where $e_1$, ..., $e_8$ designate the eight possibilities associated with the event under consideration. We take all of these outcomes as equally possible, and assign to each of them a probability of 1/8.

What is the probability of obtaining heads *at least twice* if a coin is flipped three times? In order to answer, we have to detect all favourable cases. By examining the space of events, we realize that heads occurs at least twice in four of the eight cases listed above, namely in cases $e_1$, $e_2$, $e_3$ and $e_5$. Since we have four favourable cases out of eight, the probability that is being sought is 4/8. At a closer look, to ask what is the probability of obtaining heads at least twice amounts to asking what is the probability of obtaining the disjunction ($e_1$, or $e_2$, or $e_3$, or $e_5$). We further observe that the fraction 4/8 was obtained by summing up four addenda, each with probability 1/8. These considerations lead us to formulate the following *addition* rule:

$$p (A \lor B) = p (A) + p (B) \qquad\qquad (3)$$

where the symbol '$\lor$' stands for (non-exclusive) disjunction. The rule expresses the fundamental property of *additivity*, which allows us to calculate the probability of complex events of the disjunctive kind. Statements (1), (2) and (3) taken together express the fundamental properties of probability, from which all other properties can be derived.

It should be noticed that the events considered are such that they cannot occur simultaneously. In other words, they are *mutually exclusive* in the sense that when flipping a coin three times we can have one and only one of the possibilities included in the space of events; there is no overlap between the eight events of such a space. Obviously, not all events are of this kind. Let us take a deck of bridge cards, and let $A$ stand for 'drawing a queen' and $B$ for 'drawing a card of diamonds'. These two events are not mutually exclusive, for there is one card, namely the queen of diamonds, which has both of these properties. If we asked what is the probability of obtaining a queen *or* a card of diamonds by drawing one card out of the deck, one should not simply sum up the probability of obtaining a queen and that of obtaining a card of diamonds, because, in so doing, the queen of diamonds would be calculated twice. We then come to the following formulation of the addition rule:

$$p (A \lor B) = p (A) + p (B) - (A \cdot B)$$

where the symbol '·' stands for conjunction. This can be seen as a more general formulation of the addition rule, which reduces to the former special formulation in case $(A \cdot B)$ equals 0, namely in the case of mutually exclusive events. By means of this rule, we can calculate the probability of obtaining a queen or a card of diamonds in one draw. It amounts to $4/52 + 13/52 - 1/52 = 16/52 = 4/13$.

Shortly, we shall see how to calculate the probability of two events happening simultaneously, but before doing that we need to consider the notion of *conditional* probability. Let us go back to the coin tossed three times. Suppose that, *after* the coin has been tossed, we are given the information that *fewer than* two heads have been obtained, and that we ask what is the probability that the outcome obtained was $e_8$ – namely no heads at all – conditional on the given information. Such information reduces the space of events to a subset of the initial one. The new space includes only events $e_4$, $e_6$, $e_7$ and $e_8$. Of these possible events, only one, namely event $e_8$, is favourable to the outcome whose probability is being sought, which is therefore equal to 1/4. As suggested by this example, the notion of conditional probability is related to the updating of probability in the light of new information. Conditional probability will be denoted by $p (A \mid B)$, to mean 'the probability of $A$ given $B$', or 'the probability of $A$ conditional on $B$'.

Conditional probability allows us to introduce the notion of *independence*. Two events are independent of one another when the occurrence of the first does not influence the probability of the second, and vice-versa. Using conditional probabilities, we say that events $A$ and $B$ are independent if and only if

$$p (A \mid B) = p (A) \qquad \text{and} \qquad p (B \mid A) = p (B).$$

In other words, the information that $B$ has happened does not change the probability of the event $A$, and conversely the information that $A$ has happened does not change the probability of the event $B$. In case the two events $A$ and $B$ are not independent, we have that $p (A \mid B) \neq p (A)$ and $p (B \mid A) \neq p (B)$. We then say that the occurrence of $B$ is *relevant* to that of $A$, and vice-versa. A simple inequality between the two probabilities defines a notion of *pure* relevance, whereas if the occurrence of an event, say $B$, increases the probability of the occurrence of $A$, we say that $B$ is *positively* relevant to $A$. In that case we have: $p (A \mid B) > p (A)$. If instead $B$ decreases the probability of the occurrence of $A$, we say that $B$ is *negatively* relevant to $A$. We then have:

$p\,(A\mid B) < p\,(A)$.

Let us go back to the coin tossed three times, and imagine that this experiment is repeated twice. The space of events becomes:

| | | | |
|---|---|---|---|
| 1. HHH | $e_1$ | 1'. HHH | $e_1{}'$ |
| 2. HHT | $e_2$ | 2'. HHT | $e_2{}'$ |
| 3. HTH | $e_3$ | 3'. HTH | $e_3{}'$ |
| 4. HTT | $e_4$ | 4'. HTT | $e_4{}'$ |
| 5. THH | $e_5$ | 5'. THH | $e_5{}'$ |
| 6. THT | $e_6$ | 6'. THT | $e_6{}'$ |
| 7. TTH | $e_7$ | 7'. TTH | $e_7{}'$ |
| 8. TTT | $e_8$ | 8'. TTT | $e_8{}'$. |

Clearly, the outcomes obtained in the first tosses do not influence those of the repeated experiment, so that these are independent. We now ask what is the probability of obtaining heads both in the first *and* second experiments. This is given by the *multiplication* rule:

$$p\,(A \cdot B) = p\,(A) \times p\,(B)$$

for the case of independent events. The answer to our question is then $1/8 \times 1/8 = 1/64$.

In case the events considered are not independent, the multiplication rule takes the general formulation

$$p\,(A \cdot B) = p\,(A) \times p\,(B \mid A) = p\,(B) \times p\,(A \mid B).$$

To exemplify the case of non-independent events, we ask what is the probability of throwing a king out of a deck of cards twice, assuming that after the first draw the card is *not* put back into the deck. The probability of drawing a king on the first trial is 4/52, since there are four kings in a deck of 52 cards. The probability of drawing a king on the second trial, *given* that one king has already been drawn is clearly 3/51, therefore the probability which is being sought is $4/52 \times 3/51 = 1/221$. Notice that if the first card is put back into the deck after the first draw, the two draws become independent, because the composition of the deck is not altered. In such a case, the probability of drawing two kings in two successive trials is $4/52 \times 4/52 = 1/169$.

This exemplifies what happens with *sampling*. Sampling can be done with or without replacement, depending on whether each element drawn from a population is put back or not after it has been chosen to be part of a sample. When it is done with replacement, sampling is called *random*. In this case all throws are independent, and each member of the population has the same probability of being chosen to be part of the sample[2]. If sampling is done without replacement, independence does not hold. The significance of this distinction, however, stands in connection with a finite population – the more so if it is not very large.

It should be noticed that the multiplication rule gives us an immediate way of calculating conditional probability. By that rule we have that, if $p(B) > 0$,

$$p(A \mid B) = p(A \cdot B) / p(B)$$

which can be taken as a definition of conditional probability.

Obviously, much more can be derived within the probability calculus. For instance, one can state a *negation* rule, which allows to consider the probability of the occurrence of an event via the probability of its non-occurrence:

$$p(\sim A) = 1 - p(A)$$

where '$\sim A$' stands for the negation of $A$. Suppose that we ask what is the probability of not getting a 6 when throwing a die. The disjunction rule tells us that this probability equals $1 - 1/6 = 5/6$. This is a much simpler calculation than we should make if we made use of the disjunction rule, for in that case we would have to calculate the probability of getting $(1 \vee 2 \vee 3 \vee 4 \vee 5)$, namely $1/6 + 1/6 + 1/6 + 1/6 + 1/6 = 5/6$.

In the next section we will examine a very important result that can be derived from the basic laws of probability, namely Bayes' rule.

Note that instead of starting with absolute probability, as in this chapter, one could as well take the notion of conditional probability as primitive. In that case, absolute probability would have to be defined in terms of conditional probability, which is usually done by conditioning on a tautological proposition. Accordingly, absolute probability would be defined as follows:

$$p(A \mid T)$$

where $T$ stands for a tautology. Taking conditional probability as primitive immediately reflects the fact that probability is always relative to a body of evidence, or information. Many authors, including some of those whose ideas are discussed in this book, like John Maynard Keynes, Rudolf Carnap, Harold Jeffreys, Hans Reichenbach, Karl Popper and Bruno de Finetti, take conditional probability as fundamental. The decision to introduce the laws of probability based on absolute probability was made for the sake of simplicity.

---

[2] Strictly speaking, this claim does not hold in general, but only when the composition of the population is known. If it is unknown, information on it increases with every throw which is made, so that throws cannot be said to be independent.

By referring probabilities to propositions describing events we were able to use the operations of disjunction, conjunction and negation, as is usually done in logic. However, the properties of probabilities can also be introduced with reference to a *sample space* that we call *S*, and define as the set of all atomic events, in which case an event *A* would be associated with a subset of *S*, which includes the atomic events that are favourable to it. In this case, the set-theoretical tools can be used, and in place of disjunction, conjunction and negation, the operations of union, intersection and complementation are employed.

Instead of 'events' it is often useful to speak of *random*, or *statistical variables*, to refer to the quantity that characterizes the result of some operation or experiment that is undertaken, and which is subjected to a probabilistic description. For instance, 'if we wish to estimate the temperature of a given mass of gas and we are given only a list of possible values of the speeds of its molecules, then we naturally ask how often a certain speed is observed. In other words, we naturally strive to determine the *probabilities* of various possible values of the random variable of interest to us' (Gnedenko and Khinchin 1962, p. 61). The enumeration of all possible values of a random variable and of the probability of each of these values lead to specify a probability *distribution* of the given random variables. Probability distributions play an essential role within statistical inference and methodology, for 'knowing the distribution of a given random variable enables one to solve all probability problems connected with it' (*ibid.*, p. 63).

Random variables can be discrete or continuous. When continuous random variables describing infinite processes are adopted, the theory of probability obviously requires more powerful tools than those adopted here. Such tools are provided by measure theory, whose application to probability was extensively studied at the turn of the nineteenth century and in the first half of the twentieth century by a number of mathematicians, including Émile Borel, Maurice Fréchet, Harald Cramér, Andrej Andreevich Markov, Andrej Nikolaevich Kolmogorov, Richard von Mises, Joseph L. Doob and many others[3].

---

[3] For an excellent treatise, which makes clear to what extent probability theory can be done without measure theory and for what purposes the measure-theoretic machinery is instead required, see Feller (1950, 1966). Volume I deals with discrete probabilities; volume II deals with infinite processes and continuous probabilities. For an historical account of the contribution of measure theory to modern probability see von Plato (1994).

## 2.2 Bayes' rule

A preliminary consideration that can be useful in order to understand Bayes' rule is that the conditional probability $p(A \mid B)$ and its converse $p(B \mid A)$ can be utterly different. This is shown in a simple example by Skyrms: 'The probability that Ezekiel is an ape, given that he is a gorilla, is 1. But the probability that Ezekiel is a gorilla, given that he is an ape, is less than 1' (Skyrms 1966, 2000, p. 130). Bayes' rule focuses on the relationship between two conditional probabilities, namely the probability of a hypothesis given a piece of evidence – for instance, an experimental result of some sort – and, conversely, the probability of that piece of evidence given the hypothesis at hand.

So, we now apply the notion of conditional probability to a hypothesis, symbolized as $H$, and a body of evidence, symbolized as $E$. The hypothesis in question can be of any kind: it could be a scientific law which is being conjectured, or it might regard situations encountered in everyday life. As we saw in the last section, by the multiplication rule we have that

$$p(H \mid E) = p(H \cdot E) / p(E).$$

Bayes' rule derives directly from this formula. To see this, we need to go through some simple passages. According to standard logic, $E$ is logically equivalent to $[(E \cdot H) \vee (E \cdot \sim H)]$. Substituting this expression for $E$ we have that

$$p(H \mid E) = p(H \cdot E) / p[(E \cdot H) \vee (E \cdot \sim H)].$$

Applying the disjunction rule to the denominator we obtain

$$p(H \mid E) = p(H \cdot E) / [p(E \cdot H) + p(E \cdot \sim H)].$$

We next apply the conjunction rule to both the nominator and denominator, to obtain

$$p(H \mid E) = [p(H) \times p(E \mid H)] / [p(H) \times p(E \mid H) + p(\sim H) \times p(E \mid \sim H)].$$

This is a version of Bayes' rule, taking into account only a hypothesis and its negation. A more general formulation is obtained by considering, instead of the two alternatives $H$ and $\sim H$, a set of mutually exclusive and exhaustive hypotheses $H_1, H_2, \ldots, H_n$. Since it can be proven that $E$ is equivalent to the disjunction $[(E \cdot H_1) \vee (E \cdot H_2) \vee \ldots \vee (E \cdot H_n)]$, Bayes' rule can be formulated, with reference to such a *family* of hypotheses, as follows:

$$p(H_i \mid E) = [p(H_i) \times p(E \mid H_i)] / \sum_{i=1}^{n} [p(H_i) \times p(E \mid H_i)].$$

The probability $p(H_i \mid E)$ is the *posterior* – also called *inverse*, or *final* –

probability of a certain hypothesis *given* a certain body of evidence. The rule says that this is proportional to the product of the *prior* (also called *antecedent*, or *initial*) probability of the hypothesis – calculated on the basis of background knowledge $K$, without taking into account $E$ – and the so-called *likelihood* of the piece of evidence $E$ *given* the considered hypothesis, that is on the assumption that the considered hypothesis holds. If we wanted our formulation of Bayes' rule to reflect the fact that there is always some background knowledge behind any application of the rule itself, we could make it explicit by adopting the following formulation, where $K$ is made explicit:

$$p(H_i \mid K \cdot E) = [p(H_i \mid K) \times p(E \mid K \cdot H_i)] / \sum_{i=1}^{n} [p(H_i \mid K) \times p(E \mid K \cdot H_i)].$$

Bayes' rule gives us a method of evaluating the probability of hypotheses in the light of experimental evidence. For this reason, it is commonly agreed that Bayes' rule 'tells us a great deal about the confirmation of hypotheses' (Earman and Salmon 1992, p. 71). As already pointed out in Chapter 1, the rule is also regarded as providing a method for detecting the probability of causes. The following example, borrowed from Hans Reichenbach, illustrates the use of Bayes' rule[4].

A factory that we call $A$ has three machines, that we call $B_1$, $B_2$ and $B_3$, for the production of bolts. Machine $B_1$ produces 10,000 pieces daily; machine $B_2$ produces 20,000 pieces, machine $B_3$ produces 30,000 pieces. All three machines occasionally produce faulty pieces, $C$. On the average, the rejection rates of the three machines are as follows: 4% in the case of $B_1$, 2% in the case of $B_2$, 4% in the case of $B_3$. Given a sample taken from the rejects, we ask for the probability that it was produced by each of the three machines. We calculate such a probability by means of Bayes' rule. To start with, prior probabilities, calculated on the basis of the information concerning the production of the machines, are the following:

$p(B_1 \mid A) = 10,000/60,000 = 1/6$

$p(B_2 \mid A) = 20,000/60,000 = 1/3$

$p(B_3 \mid A) = 30,000/60,000 = 1/2.$

The likelihoods are as follows:

$p(C \mid A \cdot B_1) = 4\% = 4/100$

$p(C \mid A \cdot B_2) = 2\% = 2/100$

---

[4] See Reichenbach (1935, 1949, 1971), pp. 93-4.

$p\ (C\ |\ A \cdot B_3) = 4\% = 4/100.$

Posterior probabilities are calculated as follows:

$p\ (B_1\ |\ A \cdot C) = (1/6 \times 4/100)\ /\ [(1/6 \times 4/100) + (1/3 \times 2/100) + (1/2 \times 4/100)] = 1/5 = 20\%$

$p\ (B_2\ |\ A \cdot C) = (1/3 \times 2/100)\ /\ [(1/6 \times 4/100) + (1/3 \times 2/100) + (1/2 \times 4/100)] = 1/5 = 20\%$

$p\ (B_3\ |\ A \cdot C) = (1/2 \times 4/100)\ /\ [(1/6 \times 4/100) + (1/3 \times 2/100) + (1/2 \times 4/100)] = 3/5 = 60\%.$

We therefore have a probability of 20% that a reject bolt taken at random was produced by machine $B_1$, a probability of 20% that it was produced by machine $B_2$ and a probability of 60% that it was produced by machine $B_3$. It should be noticed how the final probabilities were determined through the concurrence of the distribution of initial probabilities – which were not all equal – and the likelihoods. As observed by Reichenbach:

> Though the second machine works twice as well as the first, it is equally probable that the rejected piece originates from the second as from the first machine; this is due to the fact that the second machine produces twice as many pieces. The third machine, which supplies half of the total production, is to be assigned the probability 3/5 of having produced the reject; this probability is greater than 1/2 because one of the two other machines works more reliably (Reichenbach 1935, 1949, 1971, p. 94).

As further pointed out by Reichenbach, if one started with an equal distribution of prior probabilities, the determination of posterior probabilities would be basically determined only by the efficiency rates (likelihoods). Clearly, in the case under examination an equal distribution of prior probabilities would not be justified, as information is available, telling us that prior probabilities are different.

Knowledge of prior probabilities is crucial to the application of Bayes' rule. Borrowing Reichenbach's words once again, one can say that 'the probability of causes cannot be calculated without a knowledge of the antecedent probabilities' (*ibidem*). Sure enough, like all canons of the probability calculus, Bayes' rule does not tell us how to fix prior probabilities. The problem of determining priors is at the core of a vast debate. Bayesian statisticians are divided into 'objectivists', who claim that there are objective criteria guiding a univocal choice of prior probabilities, and 'subjectivists', who take a more liberal attitude in this connection. The problem is obviously intertwined with that of the interpretation of probability. In the following chapters we will see how various authors, including Hans Reichenbach, Harold Jeffreys and Bruno de Finetti, upholders of different interpretations of probability, take different positions on how to fix priors in view of the application of Bayes' rule.

Thomas Bayes, in the famous paper read to the Royal Society by Richard Price in 1763, introduces his rule with reference to an example. This regards a billiard table, which is

> so made and levelled, that if either of the balls $O$ or $W$ be thrown upon it, there shall be the same probability that it rests upon any one equal part of the plane as another, and that it must necessarily rest somewhere upon it (Bayes 1763, 1970, p. 140).

In this example, the physical conformation of the billiard table justifies a uniform distribution of probability of the priors. In other words, Bayes' application of his inferential method to equal prior probabilities rests on the physical characteristics ascribed to the billiard table in his example. But Bayes' text also includes a *scholium*, which contains the claim that the same assumption can be made, by analogy with the considered example, in those cases in which 'we absolutely know nothing antecedently to any trials' (*ibid.*, p. 143). In this way, Bayes' method acquires general applicability, and stands as a candidate to represent inductive inference in a probabilistic fashion.

The assumption of equal prior probabilities, which Bayes justifies by analogy with the example of the billiard table, was to be elevated to a general principle by Laplace, whose definition of probability was adopted throughout the first section of this chapter. As argued in the next chapter, such an assumption became the cornerstone of Laplace's classical interpretation of probability. Embedded in Laplace's viewpoint, Bayes' rule held undisputed for a long time, until Laplace's perspective itself underwent a crisis, and other approaches to probability took shape. In the debate that followed, the place of Bayes' inductive method within the whole of statistics became the object of a major controversy, which is still underway. To be sure, no one calls into question the validity of Bayes' rule, which is a result of the probability calculus, and follows from one of the fundamental principles of mathematical probability. What is being disputed is rather its place and role within statistical methodology. While Bayesian statisticians claim that the method reflected by Bayes' rule covers the whole of statistical methodology, other schools disagree with this tenet[5].

As we will see, Bayesian method is typically embraced by subjectivists like Bruno de Finetti, but proponents of other interpretations of probability have also assigned it a privileged role. This is the case of Hans Reichenbach, upholder of a frequentist view of probability, who repeatedly stressed the fruitfulness of Bayesian method not only within sophisticated statistical methodology, but in everyday life as well:

> The range of application of Bayes' rule is extremely wide, because nearly all inquiries into the causes of observed facts are performed in terms of this rule. The *method of indirect evidence*, as this form of inquiry is called, consists of inferences that on closer analysis can be shown to follow the structure of the rule of Bayes. The physician's inferences, leading from the observed symptoms to the diagnosis of a specified disease, are of this type; so are the inferences of the historian determining the historical events that must be assumed for the explanation of recorded observations; and, likewise, the inferences of the detective concluding criminal actions from inconspicuous observable data. In many instances the use of probability relations is not manifest because the probabilities occurring have either very high or very low values. Thus, when a corpse is found, it is virtually certain that a murder has been committed; and a fingerprint on the handle of a pistol may be considered as strict evidence for the assumption that a certain person $X$ has fired the pistol. That even in such cases the inference has the structure of Bayes' rule is often seen from the fact that appraisals of the antecedent probabilities are made. Thus an inquiry by the detective into the motives of a crime is an attempt to estimate the antecedent probabilities of the case, namely, the probability of a certain person committing a crime of this kind, irrespective of the observed incriminating data. Similarly, the general inductive inference from observational data to the validity of a given scientific theory must be regarded as an inference in terms of Bayes' rule (Reichenbach 1935, English edition 1949, 1971, pp. 94-5).

---

[5] Two articles of the *Encyclopedia of Statistical Sciences*, edited by Samuel Kotz and Norman Johnson, might be useful for further reading: Lindley (1982), containing an accessible account of Bayesian inference together with various arguments in its favour, and Good (1988), where Bayesian method is compared with other approaches to statistical methodology. For an overview of the Bayesian approach see Howson and Urbach (1989, 1993). See also Swinburne, ed. (2002).

## 2.3 Kolmogorov's axiomatization

Andrej Nikolaevich Kolmogorov was born in 1903 in Tombov, where his mother stopped on a journey to the Crimea, and died in Moscow in 1987. After his mother died in childbirth, he was raised by his aunt, and by the time he enrolled at the university of Moscow in 1920, he had developed an interest in Russian history. Soon after he started his career as a student, he became extremely fond of mathematics, and was able, within a short time, to obtain original results. In 1924, he started a fruitful collaboration with Alexandr Yakovlevich Khinchin, with whom he did important work on convergent series and the law of large numbers. Kolmogorov gave substantial contributions to various branches of probability and statistics, and is considered one of the most influential mathematicians of the last century. He was for many years professor of mathematics at Moscow State University, and member of prestigious institutions all over the world.

In 1933, Kolmogorov's famous monograph *Grundbegriffe der Wahrscheinlichkeitsrechnung* was published, where an axiomatic approach to probability is spelled out. This is inspired by the conviction that

> The theory of probability, as a mathematical discipline, can and should be developed from axioms in exactly the same way as Geometry and Algebra. This means that after we have defined the elements to be studied and their basic relations, and have stated the axioms by which these relations are to be governed, all further exposition must be based exclusively on these axioms, independent of the usual concrete meaning of these elements and their relations (Kolmogorov 1933, English edition 1950, p. 1).

The axiomatization was meant to shed light on the basic mathematical properties of probability, and to draw a distinction between probability's formal features and the meaning it receives in practical situations. As a result of Kolmogorov's axiomatization, 'probability theory acquired an equitable position among other mathematical disciplines' (Maistrov 1974, p. 262).

Kolmogorov takes the notion of probability as primitive, and proceeds to axiomatize it within the mathematical framework of measure theory. The objects to which probability is applied are *elementary events*. Like that of probability, the concept of event is taken as primitive. For the sake of axiomatization, events are treated as sets and dealt with in a set-theoretical fashion. Kolmogorov maintains that there is no need to give a univocal definition of events, precisely as the basic notions of geometry, such as 'point' and 'line', are not defined in axiomatic geometry. He proceeds to give axioms for the mathematical properties of probability functions, defined in an absolute (unconditional) sense. Kolmogorov's axioms basically reflect the properties described in Section 2.1 by means of statements (1) – (3), albeit expressed within the language of set-theory. He deals first with elementary probabilities – referring to a finite number of events – then with infinite probability fields – referring to an infinite number of random events. Random variables and distributions are then introduced, together with a theory of mathematical expectation, which is considered 'a major success of Kolmogorov's approach' (Dawid 1994, p. 1405). Kolmogorov's monograph ends with a chapter on the law of large numbers. Kolmogorov's axiomatization met with a wide consensus, and came to be regarded as the 'standard' calculus of probabilities. Naturally, it also raised some objections. A widely debated issue regards the notion of complete, or countable, additivity introduced by Kolmogorov, whereas a number of authors, such as Bruno de Finetti, have opted for a finite version of additivity[6].

While concentrating on the mathematical properties of probability, Kolmogorov's axiomatization ignores the meaning of probability, as determined in practical situations. By doing so, it also keeps off the problem of the interpretation of probability. In Kolmogorov's words:

> Every axiomatic (abstract) theory admits, as is well known, of an unlimited number of concrete interpretations besides those from which it was derived. Thus we find applications in fields of science which have no relation to the concepts of random event and of probability in the precise meaning of these words (Kolmogorov 1933, English edition 1950, p. 1).

This is perfectly in tune with the remark made at the beginning of this chapter to the effect that probability calculus is not concerned with the determination of prior probabilities, which is left to the various interpretations of probability. As the following chapters will illustrate, every interpretation embodies a particular definition of probability, together with a method of fixing priors. To be sure, all probability functions have to obey the laws of probability calculus, which therefore work as constraints of admissibility, to be fulfilled by all interpretations of probability. In other words, according to all interpretations

---

[6] For further details the reader is referred to more technical accounts, such as that of von Plato (1994).

probability has to be a real number in the interval 0–1, and to obey the rules of addition and multiplication.

As to the interpretation of probability, Kolmogorov takes sides with the frequency view[7], though his axioms do not presuppose this interpretation. As highlighted in Chapter 4, the most widely held version of frequentism in Kolmogorov's times was that of Richard von Mises, whose approach does not regard probability as a primitive notion, but defines it in terms of other concepts[8]. For this reason, Kolmogorov points out that his own approach strays from that of von Mises, and adds that their perspectives serve different purposes. Whereas Kolmogorov wants to axiomatize the formal properties of probability quite apart from its applications, von Mises sets himself a different task, 'namely to tie up as closely as possible the mathematical theory with the empirical development of the theory of probability' (*ibid.*, p. 2).

The decisive merit of Kolmogorov's axiomatization is precisely that of tracing a clear-cut boundary between the mathematical properties of probability and its interpretations, keeping the mathematical notion of probability separate from its applications, as well as from its philosophical aspects. There is no doubt that Kolmogorov's work shed much light on the notion of probability, and marked a peak in the study of this notion. In particular – and most important from the point of view of this book – the major lesson to be drawn from Kolmogorov's axiomatization is that the mathematical features of probability can and should be kept separate both from its applications and the foundational and philosophical issues connected with it. This makes room for an autonomous field of enquiry, specifically concerned with the philosophical aspects of probability, involving in the first place its interpretation. Borrowing Feller's words, one can say that

> A philosophical discussion of the foundations of probability must be divorced from the mathematical theory and its applications to the same extent as a discussion of our intuitive space concept is now divorced from geometry (though this has not always been so) (Feller 1950, p. 3).

Before we embark on a survey of the position upheld by a number of authors on the philosophy of probability, it must be stressed that Kolmogorov's axiomatization, though the best known, is by no means the only one, nor was it the first to be worked out. As a matter of fact, a number of alternative

---

[7] See von Plato (1994), Chapter 7.
[8] These are the so-called 'collectives' to be described in Chapter 4.

axiomatizations have been put forward. This holds especially for conditional probability[9]. Karl Popper deserves special mention in this connection. In 'A Set of Independent Axioms for Probability' (1938), reprinted as 'Appendix *ii' in the English edition of *Logik der Forschung*, he set forth an axiomatization of conditional probability, aimed at developing 'a formal theory which would not depend upon any particular choice of an interpretation' (Popper 1934, English edition 1959, 1968, p. 318). A number of different axiomatizations for conditional probability are available in the literature, notably that of Rudolf Carnap, which underwent various revisions in his writings. A discussion of the axiomatizations of probability and of their interconnections would require a level of technicality that falls beyond the scope of this book[10].

---

[9] See Mazurkiewicz (1932) for an early axiomatization of conditional probability.

[10] For an extensive survey of the various axiomatizations of both absolute and conditional probability that also clarifies the relation in which they stand to one another, see Roeper and Leblanc (1999). I warmly thank an anonymous referee for calling my attention to this book.

# 3

# *The classical interpretation*

## 3.1 Laplace and the Principle of insufficient reason

Determinism

The 'classical interpretation' is usually construed as the interpretation of probability developed at the turn of the nineteenth century by the mathematician, physicist and astronomer Pierre Simon de Laplace. One of the most outstanding figures in the history of science, Laplace (1749-1827) was born in Beaumont-en-Auge, Normandy, and was educated first in Caen and then in Paris, where he made great progress in mathematics, under the guidance of Jean d'Alembert. As early as in 1773, he was elected to the Academy of Sciences, to become one of the most prominent and revered scientists of his age, in France and elsewhere. After the Revolution, Laplace was engaged in a number of public enterprises, including the reform of the metric system and the census. During the Terror and the Jacobin dictatorship, Laplace left Paris with his wife and two children, but a few years later was able to resume his public life and influence. Under Napoleon, Laplace was first nominated minister of the interior – a post he held for only six months – and then chancellor of the senate. In 1805 he was appointed to the Legion of Honour, in 1816 elected to the Académie Française, and in 1817 Louis XVIII made him a marquis.

Laplace's superb systematization of Newton's mechanics – reflected in his *Traité de mécanique céleste* (1798-1825) – won him the title of 'the Newton of France', as Poisson called him at his funeral[1]. Laplace took Newtonian

---

[1] See Hahn (1967).

mechanics as the pillar on which the entire edifice of human knowledge should be made to rest. On Newtonian mechanics he claimed to ground not only the physics of terrestrial and celestial bodies, but also a cosmology, including a hypothesis on the origin of the universe. The latter is often mentioned as the 'Kant-Laplace hypothesis', on account of its similarity with the German philosopher's cosmological theory. With Laplace's work, Newtonian mechanics reached its peak as a fully-fledged paradigm of knowledge, strong enough to represent our universe in a true and incontrovertible fashion. A key ingredient of this paradigm is determinism, which Laplace considered an essential component of Newtonian theory[2].

In this perspective, probability can only have an epistemic meaning, resulting from the intrinsic limitation of human knowledge. Laplace devotes two famous essays to probability: *Théorie analytique des probabilités* (1812) and *Essai philosophique sur les probabilités* (1814)[3]. The latter opens with the introductory statement that

> The most important questions of life [...] are indeed, for the most part, only problems in probability. One may even say, strictly speaking, that almost all our knowledge is only probable; and in the small number of things that we are able to know with certainty, in the mathematical sciences themselves, the principal means of arriving at the truth – induction and analogy – are based on probabilities, so that the whole system of human knowledge is tied up with the theory set out in this essay (Laplace 1814, English edition 1995, p. 1).

This probabilism deepens its roots in an explicit recognition of man's incapability of grasping the causal connections holding between natural events:

> All events, even those that on account of their rarity seem not to obey the great laws of nature, are as necessary a consequence of these laws as the revolutions of the sun. Ignorant of the bonds that link them to the entire system of the universe, we have made them to depend on final causes, or on chance, according as they occur and succeed each other in a regular fashion, or without apparent order. But these fancied causes have been successively moved back as the boundaries of our knowledge have expanded, and they vanish entirely in the face of a sound philosophy, which sees in them only the expression of our ignorance of

---

[2] For a scientific portrait of Laplace see Gillispie (1997).
[3] The works of Laplace are collected in Laplace (1878-1912).

the true causes (*ibid.* p. 2).

The 'sound philosophy' to which Laplace refers is determinism, namely the doctrine, based on the 'principle of sufficient reason', that causality is the overall rule governing the universe, and 'The connection between present and preceding events is based on the evident principle that a thing cannot come into existence without there being a cause to produce it' (*ibidem*). Accordingly, the history of the universe is a long causal chain, where every state is determined by the preceding one, and in turn determines the subsequent one, in ways that can be described by the laws of Newtonian mechanics. Determinism, causality and predictability of events are the inextricable features of the same epistemological paradigm of classical mechanicism, whose strength derives from the precision, rigour and success of Newtonian mechanics.

Due to its limitation, the human mind is incapable of grasping all of the connections of this causal network. But one can conceive of a superior intelligence, able to apprehend such a network in every detail:

> An intelligence that, at a given instant, could comprehend all the forces by which nature is animated and the respective situation of the beings that make it up, if moreover it were vast enough to submit these data to analysis, would encompass in the same formula the movements of the greatest bodies of the universe and those of the lightest atoms. For such an intelligence nothing would be uncertain, and the future, like the past, would be open to its eyes (*ibidem*).

It should not pass unnoticed that it is by means of the powerful methods of mathematical analysis, that the super intelligence devised by Laplace is supposed to embrace the phenomena of the macro and micro cosmos, and arrange them in a unique framework. By developing and adopting such methods, the human mind has been able to figure out a comprehensive view of the 'system of the world'. Even though the superior intelligence will remain out of reach, the same methods will presumably enable man to proceed further and further:

> The human mind affords, in the perfection that it has been able to give to astronomy, a feeble likeness of this intelligence. Its discoveries in mechanics and in geometry, joined to the discovery of universal gravitation, have enabled it to comprehend in the same analytical expressions the past and future states of the system of the world. In applying the same method to some other objects of its knowledge, it has

succeeded in relating observed phenomena to general laws, and in anticipating those that given circumstances ought to bring to light. All these efforts in the search for truth tend to lead the mind continually towards the intelligence we have just mentioned, although it will always remain infinitely distant from this intelligence (*ibidem*).

In an attempt to broaden his knowledge, man resorts to probability, which – as already observed – is regarded by Laplace as an epistemological tool of the utmost importance, both in science and everyday life.

## The Principle of insufficient reason

Laplace says that probability is 'relative in part to [...] ignorance and in part to our knowledge' (*ibid.*, p. 3). Made possible by the fact that some of the circumstances of a certain event are known, probability is also necessary, because some other such circumstances are unknown. Though originating from human ignorance of phenomena in their wholeness, probability is based on what is known, and is always relative to the body of available knowledge. So conceived, probability is clearly epistemic.

In order to evaluate probability, Laplace suggests that one ought to focus on the possibilities which are open, with respect to the occurrence of an event:

> Suppose we know that, of three or more events, one alone must occur, but that nothing leads us to believe that one of them will happen rather than the others. In this state of indecision, it is impossible for us to say anything with certainty about their occurrence. However, it is probable that one of these events chosen at will {or at random} will not occur, because there are several equally possible cases that exclude its occurrence, while only a single one favours it (*ibid.*, pp. 3-4).

Having set out the matter in this way, Laplace goes on to define probability as 'the ratio of the number of favourable cases to that of all possible cases' (*ibid.*, p. 6), according to the statement known precisely as the *classical* definition.

The definition is grounded on the judgment that the possibilities in question are *equally* possible, in the absence of information that would lead us to believe otherwise:

> The theory of chances consists in reducing all events of the same kind to a certain number of equally possible cases, that is to say, to cases whose existence we are equally uncertain of, and in determining the number of cases favourable to the event whose probability is sought.

The ratio of this number to that of all possible cases is the measure of this probability, which is thus only a fraction whose numerator is the number of favourable cases, and whose denominator is the number of all possible cases (*ibidem*).

The assumption that all cases taken into account are equally possible is the cornerstone of Laplace's theory of probability. The stress he placed on the dependance of the judgment of equal possibility on there being no reason to believe otherwise, inspired the term of 'principle of insufficient reason', to refer to Laplace's assumption. It is also known in the literature as 'principle of indifference', after a terminology coined by Keynes for reasons that will be clarified in Chapter 6.

For the sake of determining probability values, equally possible cases are taken as equally probable. It is noteworthy that Laplace does consider 'unknown inequalities that may exist between supposedly equal chances' (*ibid.*, p. 34), which might be taken to suggest the existence of objective probabilities. This was observed by Donald Gillies, who claimed that it would 'contradict Laplace's own view that probability is just a measure of human ignorance' (Gillies 2000, p. 18). Gillies also recalled that in his mathematical work Laplace considers the case of a biased coin, having different chances of heads and tails. However, this is not so much a case of unknown chances, but rather of chances known to be unequal, a case admitted by Laplace, who calls attention to the need to verify that the considered cases are equally possible, before applying his method. According to Laplace, if there are reasons to believe that the examined cases are not equally possible, 'one must first determine their respective possibilities, the apposite appreciation of which is one of the most delicate points in the theory of chances'(*ibid.*, p. 6). The determination of the different chances to be attributed, say, to the two sides of a biased coin would be a case for counting frequencies. Building on this kind of cases, authors like Richard von Mises will turn to the frequentist interpretation of probability.

Laplace's epistemic interpretation protects his definition of probability from the charge of being circular. In fact, once probability is taken as epistemic, it stands on a different ground from the possibility of the occurrence of events. Moreover, as remarked by the historian of probability Stephen Stigler, Laplace's assumption of uniform prior probabilities

was not a blind metaphysical assumption that whatever was unknown was necessarily equally likely to be any of its possible values. Rather, it was an implicit assumption that for ease of analysis the problem had been specified in such a manner that this principle of insufficient reason was reasonable and that, if such were not the case, other assumptions or other prior specifications would be called for. [...] Thus nonuniform prior distributions were allowed but unnecessary: the analysis for uniform prior distributions was already sufficiently general to encompass all cases, at least for the large sample problems Laplace had in mind (Stigler 1986, p. 135).

## The Rule of succession

Having defined probability, Laplace proceeds to set forth the basic principles of the calculus of probabilities. After specifying the laws to calculate compound events, he addresses the 'probability of causes', and enunciates a principle which essentially amounts to Bayes' rule. As Laplace puts it:

> The greater the probability of an observed event given any one of a number of causes to which that event may be attributed, the greater the likelihood of that cause {given that event}. The probability of the existence of any one of these causes {given the event} is thus a fraction whose numerator is the probability of the event given the cause, and whose denominator is the sum of similar probabilities, summed over all causes. If these various causes are not equally probable *a priori*, it is necessary, instead of the probability of the event given each cause, to use the product of this probability and that of the cause itself. This is the fundamental principle of that branch of the analysis of chance that consists of reasoning *a posteriori* from events to causes (Laplace 1814, English edition 1995, p. 9).

As one can see, in formulating this principle Laplace considers the case in which 'the causes' are not equally probable, but adds that 'when the probability of a simple event is unknown, one may suppose that it is equally likely to take on any value from zero to one' (*ibid.*, p. 10).

Under the assumption of equally likely causes, Laplace derives from the principle outlined above that 'when an event has happened any number of times running, the probability that it will happen again next time is equal to this number increased by 1, divided by the same number increased by 2' (*ibid.*, p. 11). Here Laplace states the famous method of inference later labelled by John Venn as the *rule of succession*, often called in the literature

'Laplace's rule'[4]. In the case of two alternatives – like 'occurrence' and 'non occurrence' – this rule allows us to infer the probability of an event, from the fact that the same event has been observed to happen in a given number of cases. In a general formulation, the rule says that if $m$ is the number of observed positive cases, and $n$ that of negative cases, the probability that the next case to be observed is positive equals

$(m + 1) / (m + n + 2)$.

If no negative cases have been observed, or one wants to calculate the probability of an event, after having observed that it has happened '$m$ times in a succession', the formula is reduced to

$(m + 1) / (m + 2)$.

The example used by Laplace to illustrate this method is in the tradition of Price, and was to raise extensive discussion in subsequent writings on probability. It regards the probability that the sun will rise tomorrow. In Laplace's words: 'if we place the dawn of history at 5,000 years before the present date, we have 1,826,213 days on which the sun has constantly risen in each 24 hour period. We may therefore lay odds of 1,826,214 to 1 that it will rise again tomorrow' (*ibid.*, p. 11).

Laplace's rule of succession is based on the assumptions of the equiprobability of priors and the independence of trials, conditional to a given parameter – like the composition of an urn, or the ratio of the number of favourable cases to that of all possible cases. The authors who later worked on probabilistic inference in the tradition of Bayes and Laplace – especially William Ernest Johnson, Rudolf Carnap and Bruno de Finetti – eventually turned to the weaker assumptions of exchangeability[5].

As we will see in more detail, Laplace's theory of probability, particularly the principle of insufficient reason, has been the object of much discussion and criticism. However, Patrick Suppes observed, it should not be forgotten that

> even though the classical definition of probability is vague and difficult to apply, it continues to be important because it is, in effect, the definition of probability that is used in many important applications in which the guiding principles of symmetry are clear and generally well agreed upon. It is the classical definition of probability that is used to

---

[4] The first formulation of this method is contained in Laplace (1774).
[5] For an excellent overview of this tradition see Zabell (1989).

make standard calculations in games of chance (Suppes 2002, p. 167).

An exposition of Laplace's contribution to probability cannot fail to mention what Stigler describes as 'Laplace's major result in probability theory' (Stigler 1986, p. 136), namely his 'central limit theorem', by which Laplace makes further progress in the analysis of direct probability, in the tradition of Jakob Bernoulli and de Moivre, then carried on by Poisson. As described by Stigler, Laplace's result generalizes de Moivre's theorem by saying that 'any sum or mean (not merely the total number of successes in $n$ trials) will, if the number of terms is large, be approximately normally distributed' (*ibidem*).

## Expectation and certainty

Laplace concludes his work on probability by introducing mathematical expectation, defined in general terms as 'the product of the sum expected and the probability of getting it' (Laplace 1814, English edition 1995, p. 11). Having stated the basic principles of mathematical expectation, Laplace recommends that

> In the conduct of life one ought always to see to it that the product of an expected gain and its probability is at least equal to the similar product relative to the loss. But in order to achieve this it is necessary accurately to estimate the gains, the losses, and their respective probabilities. This in turn calls for great precision of mind, a nice judgment, and wide experience in worldly affairs. It is necessary to know how to guard oneself against prejudice, against illusions of fear and hope, and against those treacherous notions of success {or wealth} and happiness with which most men lull their *amour-propre* (*ibid.*, p. 12).

After a brief mention of the Saint Petersburg paradox, Laplace addresses the problem of the discrepancy between the relative, or moral value of a certain good, and its nominal value, namely that obtained through calculation. Laplace describes the situation by saying that 'the moral advantage that a gain procures for us is not proportional to that gain', but rather 'depends on a thousand circumstances that are often very difficult to describe, but of which the most general and the most important is that of one's wealth' (*ibid.*, p. 13). On account of these difficulties, Laplace proposes to adopt a notion of *moral expectation*, borrowed from the work of Daniel Bernoulli.

Laplace then discusses at length the applications of probability to both the natural and moral sciences. Following the tradition of d'Alembert and Condorcet, Laplace was firmly convinced of the usefulness of probability in all fields of knowledge, including testimonies, elections and decisions of assemblies and judicial decisions, all fields where he deemed probability had been applied with success. Indeed, one can say that 'Laplace considered himself that very "philosophical mathematician" which d'Alembert and Condorcet had described' (Hahn 1967, p. 16).

Towards the end of his *Essai philosophique*, Laplace deals with the means for approaching the truth, and assigns a crucial role in this connection to induction and analogy. To be sure, a correct and fruitful use of these methods has to be grounded in empirical observations, and backed by experience. As Laplace puts it:

> Induction, analogy, hypotheses founded on facts and continually corrected by new observations, a fortunate intuition, given by nature and strengthened by numerous comparisons of the things it suggests with experience – these are the principal means of arriving at the truth (Laplace 1814, English edition 1995, p. 112).

Laplace's empiricism clearly emerges in this section of the essay, where he reaffirms that experience is the foundation of the whole edifice of knowledge: hypotheses originate from experience, and can only be amended by comparison with experience. This goes hand in hand with Laplace's unshakable faith in Newtonian mechanicism:

> The surest method to guide us in the search for truth consists in proceeding inductively from phenomena to laws, and from laws to forces. Laws are relationships that connect particular phenomena to one another; when the general principle of the forces from which they are derived has been made known by them, it may be verified, either by direct experiments, whenever that is possible, or by our investigating whether it agrees with known phenomena. And if by a rigorous analysis one finds that they all follow from this principle, down to the smallest detail, and if moreover they are very varied and very numerous, then science attains the highest degree of certainty and of perfection possible. This is what has happened to astronomy by the discovery of universal gravitation (*ibid.*, p. 116).

The mechanistic paradigm is then ascribed an important heuristic role. If his innate limits prevent man from obtaining full and certain knowledge of phenomena, probability – combined with induction and analogy, and guided by intuition – represents a reliable tool to approach the truth.

## 3.2 Problems of the classical definition

Laplace's work represents the synthesis of seventeenth and eighteenth century studies on probability, and greatly influenced subsequent research. It also aroused a vast debate, that cannot be summarized in this context. As we shall see in the next chapters, the progressive crumbling of the classical theory gave way to the other interpretations of probability. Of these, the first to take shape was the frequentist interpretation. By embracing an empirical notion of probability and a way of measuring probability based on frequency, frequentists soon got rid of the principle of insufficient reason. Achieving this goal was more complicated for those who followed Laplace in taking probability as an epistemic notion. Among them, the principle of insufficient reason, though the subject of stern criticism, held the stage until the work of authors like Johnson, de Finetti and Carnap, whose work will be considered in the following chapters.

While various objections to Laplace's position will be discussed in some detail in the following pages, the main charge moved by frequentists can be summarised by saying that there are many circumstances, in which it does not seem possible to determine the set of all possible cases, and even less the set of 'equally likely' cases. Both in science and everyday life, we face situations where, instead of looking for possible cases, we count the frequency with which events take place in order to calculate probability. Cases in point are the probability of a biased coin falling on either sides, or the probability that a given individual will die within a year. In view of this, frequentists invoke abandoning what Ernest Nagel calls 'a form of a priori rationalism' (Nagel 1939a, 1969, p. 48). The main charge moved by the upholders of an epistemic view of probability, logicists in particular, is that the principle of insufficient reason is really an equidistribution of ignorance, and ends up by grounding probability on ignorance, rather than knowledge. This is considered not only arbitrary, but too restrictive, as it hinders learning from experience.

In addition to all this, the classical interpretation of probability suffers from a problem of applicability when applied to problems involving an infinite number of possible cases. This difficulty is at the basis of the 'Bertrand's paradox', after the French mathematician Joseph Bertrand (1822-1900), professor of analysis at the École Polytechnique in Paris and later elected to the Collège de France. Bertrand was a distinguished mathematician, as well as a writer of articles in popular science journals, like the *Journal des savantes* – which he edited from 1865 to his death, and the *Revue des deux mondes*. His *Calcul des probabilités* (1888) contains the exposition of the problem usually associated in the literature with his name, after Henri Poincaré coined the term in Chapter VII of his own *Calcul des probabilités*, published in 1912.

'Bertrand's paradoxes' cover a whole cluster of problems involving an infinity of possibilities[6]. In such cases, the classical definition of probability, together with the principle of indifference, leads to contradictory results, that is, incompatible probability assignments. As Bertrand puts it:

> the infinite is not a number, one must not introduce it into arguments without explication. The illusory precision of words might give rise to contradictions. To choose *at random* among an infinite number of possible cases is not enough as a precept.
>
> One could ask, for instance, what is the probability that a number, whole or fractional, commensurable or incommensurable, chosen *at random* between 0 and 100 be greater than 50. The answer seems evident: the number of favourable cases is half that of possible cases. The probability is 1/2.
>
> Then, instead of a number, one could take its square. If the number lies between 50 and 100, its square will lie between 2,500 and 10,000.
>
> The probability that a number chosen *at random* between 0 and 10,000 be greater that 2,500 seems to be evident: the number of favourable cases is three quarters of the number of possible cases. The probability is 3/4.
>
> The two problems are identical. Whence comes the difference between the answers? The statements lack precision.
>
> Contradictions of this kind can be infinitely multiplied (Bertrand 1888, p. 4).

There follows the description of other problems, also discussed by Poincaré. The first regards the probability that the chord of a circle be shorter than the side of the equilateral triangle inscribed in it. A further problem is that of determining the probability that a plane chosen *at random* in space will form

---

[6] Stigler credits Francis Ysidro Edgeworth for having called attention, in an essay of 1885, to the problems raised by the principle of insufficient reason in connection with cases involving an infinite number of possibilities. See Stigler (1986), p. 127.

with the horizon an angle smaller than $\pi/4$. Finally, Bertrand sets the problem of calculating the probability that, given two points fixed at random on the surface of a sphere, the distance between them be less that 10'. In all such cases, he argues that one faces ill-posed questions, bound to lead to contradictory results.

Clearly, there are infinitely many problems falling under the scope of 'Bertrand's paradoxes'[7]. Let us borrow a very simple example, taken from everyday life, from Wesley Salmon:

> Suppose we know that a car has taken between one and two minutes to traverse one mile, and we know nothing further about the amount of time taken. Applying the principle of indifference, we conclude that the probability that the time was between one and one-and-one-half minutes equals the probability that it was between one-and-one-half and two minutes. Our data can, however, be expressed in another way. We know that the *average* speed for the trip was between sixty and thirty miles per hour, but we know nothing further about the average speed. Applying the principle of indifference again, we conclude that the probability that the average speed was between sixty and forty-five miles per hour equals the probability that it was between forty-five and thirty miles per hour. Unfortunately, we have just contradicted the first result, because the time of one-and-one-half minutes corresponds to forty, not forty-five, miles per hour. (Salmon 1966, pp. 66-7).

Salmon observes that

> This example is not just an isolated case (although one contradiction ought to be enough), but it illustrates a general difficulty. A similar conflict can be manufactured out of any situation in which we have two magnitudes interdefined in such a way that there is a nonlinear relation between them (*ibidem*).

Going back to Bertrand, it is noteworthy that he is critical of a generalized application of Laplace's rule of succession. In this connection, he writes that 'almost all of the applications of this formula have been ill-founded. People have dared to look for the probability that the sun will rise tomorrow', but an assertion of this kind can only be confirmed 'through the discovery of the laws

---

[7] For a detailed discussion of the paradoxes raised by the classical interpretation of probability see Suppes (2002), Chapter 5.

of astronomy, not through the renewed success of a game of chance' (Bertrand 1888, p. 174).

Bertrand is led by his analysis to doubt the universal applicability of Laplace's method. Instead, he shares Laplace's determinism, and maintains that 'the intervention of chance in the formation of the universe is not acceptable. Everything is ruled by a law, there is no doubt about it' (*ibid.*, p. 171). But by the time Bertrand was writing his treatise on probability, determinism was already trembling, as the new developments of science were gradually undermining classical mechanics, and the view of knowledge based on it. The climax of this process can be identified with the advent of quantum mechanics at the beginning of the twentieth century. This made indeterminism look no less viable a hypothesis than determinism, and in the nineteenth century statistical mechanics and evolutionary biology were progressively opening science to probability. The analysis of mass phenomena typical of these disciplines, *ending* up with the extensive use of average values calculated on the basis of frequencies, contributed to the growth of the frequency interpretation of probability.

# 4

## The frequency interpretation

### 4.1 Robert Leslie Ellis

In the mid-nineteenth century Robert Leslie Ellis (1817-1859), a mathematician fellow of Trinity College, read two essays at the Cambridge Philosophical Society committed to a view of probability strictly linked with frequency. The first, 'On the Foundations of the Theory of Probabilities', read in 1842 and published in the *Transactions* in 1849, was followed by 'Remarks on the Fundamental Principle of the Theory of Probabilities', read in 1854 and published in 1856[1].

The opening statement of the first essay illustrates Ellis' main concern, namely to address the philosophical problem of the foundations of probability:

> The theory of probabilities is at once a metaphysical and a mathematical science. The mathematical part of it has been fully developed, while, generally speaking, its metaphysical tendencies have not received much attention (Ellis 1849, p. 1).

The use of the term 'metaphysical' should not mislead us, as in this context it simply means 'philosophical'. In tackling the foundations of probability, Ellis is guided by the conviction that one should start from an empirical perspective, apt to link probability to experience. Probability judgments are based on experience, and this is in turn based on sensations, so Ellis wants to 'call attention to the inconsistency of the theory of probabilities with any other than a *sensational* philosophy' (*ibidem*).

---

[1] See Ellis (1849) and (1856). Both essays are reprinted in Ellis (1863), containing also a 'Biographical Memoir' by Harvey Goodwin.

According to Ellis, a careful analysis of both uncertain events and the laws governing them, namely the mathematical theory of probability, suggests that probability rests ultimately on frequency. The expectation of a given result on a certain trial is of necessity associated with the experience of the frequencies exhibited by series of similar trials. As Ellis puts it:

> For myself, after giving a painful degree of attention to the point, I have been unable to sever the judgment that one event is more likely to happen than another, or that it is to be expected in preference to it, from the belief that on the long run it will occur more frequently (*ibid.*, p. 2).

Thus, the frequency with which events happen is really the kernel of probability judgments, which are not influenced at all by the varying circumstances that accompany the single occurrences of events, being instead focussed on the similarities that are detected by means of a comparison between events of the same kind. Once they are considered not in isolation, but as members of long series of similar cases, events tend to show permanent features, or 'what is called the nature of the case' (*ibid.*, p. 3). Following this line of thought, Ellis comes to 'the fundamental axiom, that on the long run, the action of fortuitous causes disappear' (*ibidem*). He takes this axiom 'to be an *a priori* truth, supplied by the mind itself, which is ever endeavouring to introduce order and regularity among the objects of its perception' (*ibidem*). In other words, this principle, on which the theory of probability ultimately rests, is an 'intuitive' principle, taken as a component of the very act of thinking about probability.

The intuitive and aprioristic character of the fundamental principle of probability should not conceal the fact that probability judgments are about empirical phenomena, more precisely about repeatable phenomena of a certain kind; they do not concern personal opinion. Furthermore, probability cannot be applied without restriction to all sorts of uncertain phenomena. Ellis is very clear in this connection:

> The results of the theory of probabilities express the number of ways in which a given event can occur, or the proportional number of times it will occur on the long run: they are not to be taken as the measure of any mental state; nor are we entitled to assume that the theory is applicable wherever a presumption exists in favour of a proposition whose truth is uncertain (*ibidem*).

In the second essay, Ellis restates the fundamental principle of the theory of probabilities as follows: 'on a long run of similar trials, every possible event tends ultimately to recur in a definite ratio of frequency' (Ellis 1856, p. 605). He then makes an attempt to clarify the meaning of the expression 'similar trials'. In order to better specify its meaning, the author resorts to the notions of *group* – or *genus* – and *species*. With respect to trials belonging to a certain group or genus, possible results represent different species. For instance, if the trial consists in throwing a die, one can say that this is a group, while the occurrences of its sides stand for distinct species.

Probability theory requires in addition that the series of trials be indefinitely long. Clearly, the very idea of an infinite series of trials raises difficulties, because no such series can ever be obtained in practice. Ellis answers that such an infinite series can be taken as an idealization, or as a realization – impossible in practice – of what is potentially included in the series. He adds that similar idealizations recur in all fields of science, where it is common to make use of concepts that stand in an imperfect, usually approximate, correspondence with facts. As an example, Ellis mentions the idea of an infinite plane, required by the first law of motion:

> The indefinitely prolonged series of trials, which enters into the ordinary statement of the fundamental principle of the theories of probabilities, is analogous to the infinite and indefinitely smooth horizontal plane, which would enable us to verify the first law of motion (*ibid.*, p. 607).

The need for idealized concepts does not regard only the infinite character of probability series, but also the similarity between the elements of the series. Obviously, natural phenomena are never identical, therefore when speaking of 'similar trials' an idealization process is performed, so that the idea of a series of similar trials is obtained by abstraction from the differences characterizing each singular case.

To Ellis, frequentism looks like the only viable option, once it is agreed that 'the principle on which the whole depends, is the necessity of recognizing the tendency of a series of trials towards regularity, as the basis of the theory of probabilities' (Ellis 1849, p. 6). According to Wesley Salmon, who has devoted great attention to the frequency interpretation of probability, Ellis' ideas 'took us to the very threshold of a frequency theory of probability' but it was John Venn who 'opened the door and led us in' (Salmon 1980b, p. 143). In other words, Ellis' work points in the right direction, and contains the main ingredients of the frequency interpretation, but it was Venn who brought the interpretation to completion[2].


## 4.2 John Venn

### Probability as limiting frequency

Born in Drypool (Hull) and member of Gonville and Caius College in Cambridge, of which he was President at the time of his death, John Venn (1834-1923) was an outstanding logician, well known for elaborating a method for testing categorical syllogisms by means of the so-called 'Venn diagrams'. In *Symbolic Logic* (1881) he also made important contributions to Boolean algebra, while *The Principles of Empirical or Inductive Logic* (1889) contains a critical discussion of Mill's methods. Though praising Mill for having developed what he regards as a tenable theory of induction, Venn objects that Mill's methods are hardly applicable to real science, where the phenomena to be investigated are far more complex than assumed by such methods. This does not mean that they are useless, for they are theoretically sound and liable to suggest fruitful directions in which research should be pursued.

As concerns probability, Venn moves a step forward in the direction pointed by Ellis, writing the first book devoted to the frequency interpretation of probability, namely *The Logic of Chance*[3]. In the preface to the first edition (1866), Venn complains that the field of probability 'has been very much abandoned to mathematicians, who as mathematicians have generally been unwilling to treat it thoroughly' (Venn 1866, 1876, p. viii). In the work of mathematicians, he continues,

> we would search in vain for anything like a critical discussion of the fundamental principles upon which its [of probability] rules rest, the class of enquiries to which it is most properly applicable, or the relation it bears to Logic and the general rules of inductive evidence (*ibidem*).

The goal Venn sets himself is that of dealing with the philosophical foundations of probability, to the purpose of both clarifying the conceptual

---

[2] See Salmon (1980b) and (1980a) for a competent account of Ellis' and Venn's contribution to the frequency interpretation of probability.

[3] This book was published in three editions, in 1866, 1876 and 1888. The second and third editions contain substantial changes with respect to the first ones.

nature of probability and making it 'interesting and intelligible to ordinary readers who have any taste in philosophy' (*ibid.*, p. xi). Venn's treatment of the matter is informal, and makes almost no use of mathematics, which occasionally results in a weakness of the exposition. In a memoir of Venn, Henry Thomas Francis reports that Venn was trained in mathematics when he was a student, but after three and a half years of such training he developed a strong reaction and even felt disgust for mathematics, which made him lose part of what he had acquired. Francis also reports that later on in life Venn admitted that that was a great mistake on his part and that he should have retained his knowledge of mathematics and utilized it for his work on probability and logic[4].

Venn recognizes that there are exceptions to the widespread tendency to reduce probability to its mathematical aspects, notably those represented by Augustus De Morgan and George Boole, who – as we shall see in Chapter 6 – embrace an epistemic notion of probability. Although sharing these authors' interest in the foundational aspects of probability, Venn strays from their approach, which he considers based on a 'conceptualist' logic, to embrace a frequentist notion of probability, similar to that of Ellis. According to Venn, probability belongs to logic, but a 'material' logic aimed at taking 'cognizance of the laws of things', not a 'conceptualist' logic, having to do with 'the laws of our own minds in thinking about things' (*ibid.*, p. x). Probability as a branch of material logic is ultimately based on empirical knowledge, and never loses reference to facts. Venn's radically empirical attitude in this connection is summarized by saying that 'our ultimate reference is always to facts. We start from them as our data, and reach them again eventually in our results whenever it is possible' (*ibid.*, p. 279). Starting from the experienced, probability leads to infer non experienced facts.

In developing his frequentist approach to probability, Venn abandons Ellis' terminology of genus and species, to use the notion of *series* instead. This is the building block of Venn's theory, which is therefore based on

> series of things or events, about the individuals of which we know but little, at least in certain respects, whilst we find a continually increasing uniformity as we take larger numbers under our notice (*ibid.*, p. 9).

Insofar as we consider events taken individually, we can only see variation and never find uniformity, but as soon as we consider collections of events, we

---

[4] See Francis (1923), p. 16.

observe similarities and detect repetitions. By counting the number of such repetitions one obtains *proportions*, or what we today call frequencies. The use of such proportions generates order out of disorder, and the rules of probability must therefore be based on proportions derived from series of repeatable events.

Like Ellis did with reference to his own theory, Venn admits that the concept of a series is an idealization, obtained by abstraction from the diversity of its components. This is regarded as a necessary step in order to make calculations applicable to objects. Plainly, this process of idealization should not overlook major differences between the objects considered, or be performed at the price of accuracy. As Venn puts it:

> In nature nearly all phenomena present themselves in a form which departs from that rigorously accurate one which scientific purposes mostly demand, so we have to introduce an imaginary series, which shall be free from any such defects. The only condition to be fulfilled is, that the substitution is to be as little arbitrary, that is, to vary from the truth as slightly as possible (*ibid.*, p. 102).

Once a series of repeatable events is specified, the calculation of probabilities is made possible. To this end, the series must be indefinitely long. In fact probabilities are calculated on the basis of 'proportions in the long run', and the latter 'must be supposed to be very long indeed, in fact never to stop' (*ibid.*, p. 146). The proportions observed in a short series of events exhibit fluctuations, but

> As we keep on taking more terms of the series we shall find the proportion still fluctuating a little, but its fluctuations will grow less. The proportion, in fact, will gradually approach towards some fixed numerical value, what mathematicians term its *limit* (*ibidem*).

With this definition, Venn lays the foundations of the frequency theory of probability, according to which probability is defined as the limiting value of frequency in infinite series of events. The infinite character of the series is again an idealization, whose adoption is unavoidable for scientific purposes. Only those series which exhibit proportions approaching a limit can be the object of probabilistic calculations and be 'made subjects of strict science' (*ibidem*).

Having settled the object and nature of probability, Venn goes on to specify the rules of probabilistic inference, namely the additive rule for mutually exclusive and non exclusive classes of events and the multiplicative rule. It is worth recalling that Venn ascribes no importance at all to the difference between direct and inverse probability. In this connection, he claims that

> the distinction between Direct and Inverse Probability must be abandoned. When the appropriate statistics are at hand the two classes of problems become identical in method of treatment, and when they are not we have no more right to extract a solution in one case than in the other (*ibid.*, p. 168).

## Criticism of the rule of succession

Venn is very critical of Laplace's rule of succession, which he feels the duty to discuss, given the prestige of its supporters, starting with Laplace himself. In the first place, Venn observes that there is some ambiguity in connection with the interpretation of the rule, which Laplace takes sometimes as a rule of inference, and at other times as a psychological principle. In Venn's words:

> I find it difficult to ascertain precisely from Laplace's Essay what his view of this Rule of Succession is. On the one hand he certainly appeals to it as a valid rule of inference, but on the other hand he enters into decidedly psychological and even physiological explanations in the latter part of his Essay. But converting any such formula into an ultimate principle we do in reality abandon it as a practical rule (*ibid.*, p. 179).

In the third edition of *The Logic of Chance*, where the chapter on the rule of succession is considerably modified, Venn adds that probably what Laplace had in mind 'was the natural history of belief rather that its subsequent justification' (Venn 1866, 1888, p. 201). This indeed looks to Venn as the only tolerable way of conceiving the rule, which would accordingly belong to psychology, rather than to logic or probability. In any case, elevating the rule to a general principle is a mistake. The rule can at best be taken as a practical device, that might prove useful in oversimplified cases.

Venn shares the critical attitude towards the rule of succession expressed by other authors, like Boole and De Morgan, but disagrees with the latter on the admission that the rule can at least furnish a *minimum* value for the amount of expectation. To this idea, he objects that there are cases in which experiencing that something has happened a few times does not provide good reasons for expecting it to happen again. As an example, he mentions the following: 'I have given a false alarm of fire on three different occasions and found the people came to help me each time' (Venn 1866, 1876, p. 180). In a case like this, using the rule of succession to evaluate the probability of a repetition of the same phenomenon would clearly prove mistaken in the long run.

Nor is it the case, for Venn, that the rule of succession should hold in those cases in which people do not know anything about the events under consideration:

> The truth and falsity of the rule cannot in any way be dependent upon the ignorance of the man who uses it. [...] We cannot fling the rule amongst mankind with the prescription attached that it is merely to be taken by the ignorant (*ibid.*, p. 181).

Venn concludes that with the rule of succession 'a formula is given to us which really does not seem fitted to put those who trust to it oftener in the right path than if they had never heard of it' (*ibid.*, pp. 181-2). The only reliable guide to our expectations is empirical evidence delivered by observed series, and it is on the latter that probability should be grounded.

## Probability and belief

While reaffirming the idea that 'probability is a science of inference about real things' (*ibid.*, p. 101), Venn strongly opposes the view that probability is a measure of belief. He devotes many pages to criticizing this conception, especially as held by Augustus De Morgan, W. Thomson and William Donkin. His main argument is that belief is formed out of a multitude of factors, so complex and dependent on time and context, as to make it inadequate to represent probability in any precise and objective way. The objective element of belief is evidence, and it is in response to evidence that belief is modified. The only acceptable way of addressing the problem of how we reason in the face of experience is therefore to be sought in the kind of evidence we acquire by empirical observation. In the case of uncertain events, evidence consists of proportions detected within series, and this is what should be put at the basis of both the theory of probability and our belief in such events. So for Venn 'the attempts, so frequently made, to found the science [of probability] on a subjective basis [...] can lead [...] to no satisfactory results' (*ibid.*, p. 135).

However, once probability is given a proper foundation in terms of series and proportions observed in the long run, it can also be used to justify belief. At the end of a long discussion on the topic, Venn claims to have given reasons against the opinion that our belief admitted of any exact apportionment like the numerical one. [...] Still, it was shown that a reasonable explanation could be given of such an expression as, 'my belief is 1/10th of certainty', though it was an explanation which pointed unmistakeably to a series of events, and ceased to be intelligible, or at any rate justifiable, when it was not viewed in such a relation to a series. In so far, then, as this explanation is adopted, we may say that our belief is in proportion to the above fraction (*ibid.*, p. 147).

This is the only sense Venn can attach to the statement that

the fraction expressive of the probability represents also the fractional part of full certainty to which our belief of the individual event amounts. Any further analysis of the matter would seem to belong to Psychology rather than Probability (*ibid.*, p. 148).

So much for the idea that probability also gives grounds for expectation, not to be confused with the tenet that probability is itself a degree of belief.

The third edition of *The Logic of Chance* includes a new chapter on 'Chance as opposed to causation and design'. With reference to the traditional opposition between chance and causation, Venn maintains that it does not arise within his own theory. In fact, according to the latter

The science of Probability makes no assumption whatever about the way in which events are brought about, whether by causation or without it. All that we undertake to do is to establish and explain a body of rules which are applicable to classes of cases in which we do not or cannot make inferences about the individuals (Venn 1866, 1888, p. 236).

Venn adds that his own theory deals with averages, and single events are taken as derivative from series. In Venn's words:

On the view here adopted we are concerned only with averages, or with the single event as deduced from an average and conceived to form one of a series. We start with the assumption, grounded on experience, that there is uniformity in this average, and, so long as this is secured to us, we can afford to be perfectly indifferent to the fate, as regards causation, of the individuals which compose the average (*ibid.*, p. 239).

The question arises whether this assumption is incompatible with the irregularity of the individuals, due to absence of causes. Venn answers that there is no incompatibility, in so far as the two beliefs coexist in the mind of people who are ready to accept that the behaviour of such things as games of dice are predictable in the long run, but unpredictable in relation to the next throw. Having argued against any opposition between chance and causation that might be of some interest to probability, Venn discusses the antithesis of chance and causation with reference to individual phenomena, where observing coincidences leads to detection of common causes. This is taken as a problem of induction, rather than probability. Venn also discusses at length the problem of free will and the distinction between chance and design, as a product of agency. But again, these problems are seen as falling outside of the scope of probability, and having to do instead with morals or theology.

The problem of religious belief is also of concern to Venn, who in 1858 took Priests orders, but in 1870 quit the Priesthood. He addresses the issue with his usual empiricist attitude in the Hulsean Lectures of 1869, published under the title *On Some of the Characteristics of Belief Scientific and Religious* (1870). Once again, Venn adopts a 'method of treatment' which 'is logical and not metaphysical', having the advantage that 'on the field of logic [...] people of the most opposite schools may meet and shake hands' (Venn 1870, p. vi). Accordingly, belief is treated 'as being founded solely upon evidence, with the implication that in the thoughtful and sound-minded it is rightfully so founded' (*ibid.*, p. vii). Venn argues that religious and scientific belief do not differ in any substantial way, being rather based on a different type of evidence.

The difference between the two kinds of belief is to be sought in the different level of complexity characterizing the evidence on which they rest. The evidence underpinning religious belief is multifarious, and deeply conditioned by social and emotional factors that create divergences of opinion between one person and another. It therefore happens that in science, but not in religious matters, when a divergence of opinion exists, an experiment can be designed in such a way that its result counts as evidence in favour of one opinion and against another. This decisive advantage of science over religion resides in the peculiar character of the empirical evidence underpinning science. As Venn puts it:

> suppose that we are confronted with a difference of opinion deliberately adopted and persistently maintained: what is to be done? [...] How is it to be remedied? [...] In science we know what our resource would be, in case such a state of things should be found to exist there. We should have to devise some experiment and see which of the conflicting opinions was correct. Of course such an 'interrogation of nature' is one to which an answer is only to be got with difficulty; when however we

have succeeded in getting it, it is generally clear and unequivocal. But in reference to religious truth any appeal to experience in the strict sense of the term is mostly out of the question (*ibid.*, pp. 33-4).

Once again we are led out of the realm of science and probability, to enter the realm of faith, as to which science is barren.

The text under discussion also contains some interesting remarks on Pascal's wager. Venn claims that reasons to act and reasons to believe should be kept separate. While the latter are based on evidence, the former must in addition take into account the possible gain and loss connected with possible courses of conduct. Pascal's wager is based precisely on the principle that in the face of uncertain events, human action should be directed towards those actions associated with a greater benefit. In this connection, Venn says:

> I do not mean that we must *believe* events to be more probably true than they really are; far from it; but we must constantly act upon the hypothesis that an event may really happen when we well know that such a contingency is very unlikely. I know, say, that the chances are a thousand to one that my house will never be burnt down; my belief in the contingency is very small, but I act upon the truth of it and insure the house (*ibid.*, pp. 116-7).

Belief can only be enhanced by bringing more evidence, and this is not done by Pascal's wager, which should not be interpreted as concerning belief, but action. Taken in this way, it represents a reasonable principle of conduct.

## 4.3 Richard von Mises and the theory of 'collectives'

### Von Mises' approach

The most perspicacious version of the frequency theory of probability was provided by Richard von Mises, who gave it a rigorous formulation and clarified its theoretical presuppositions, while fixing the boundaries of its range of application. In dealing with probability, von Mises was animated by an utterly empiricist and operationalist attitude, typical of the philosophical movement to which he adhered, namely logical empiricism. In fact Richard von Mises (1883-1953), together with his brother the well known economist Ludwig, had taken an active part in this movement since the years before the first world war, when intellectuals of various provenance, namely physicists, mathematicians, sociologists, historians and philosophers, started meeting once a week in a Viennese café to promote a new way of treating philosophy and, more generally, a new approach to intellectual life. In the Twenties, the movement developed into the 'Vienna Circle' of Moritz Schlick, Otto Neurath, Hans Hahn, Philip Frank, Rudolf Carnap and many others, known as 'logical empiricists'. Starting with a thorough analysis of the deep changes in knowledge after the revolutionary developments of science between the nineteenth and twentieth centuries, logical empiricists fostered a radical turn in philosophy towards a 'scientific conception of the world', combining a radically empirical attitude with the adoption of the powerful conceptual tools of formal logic, as worked out by authors like Bertrand Russell[5].

After graduating in engineering in Vienna and Brno, and teaching at various universities, in 1921 von Mises was appointed by the University of Berlin, where he founded the Institute for applied mathematics. There he joined Hans Reichenbach, Kurt Grelling, Carl Gustav Hempel, Walter Dubislav and others in the 'Berlin Society for empirical philosophy', counterpart of the Vienna Circle. Being of a Jewish family, von Mises decided to leave Germany and went to Istanbul, where he remained from 1933 until 1940 as professor of mathematics and government adviser. In 1940 he moved to the United States, where he spent the rest of his life in Boston, after having being appointed professor of applied mathematics at Harvard. Von Mises was a prolific writer, and left a bulk of articles and books on a number of subjects belonging to various branches of mechanics, mathematics, probability and even literature – a great admirer of the work of the poet and novelist Rainer Maria Rilke, von Mises left the world's largest collection of Rilke's works[6].

Von Mises' best known work on probability is *Wahrscheinlichkeit, Statistik und Wahrheit* (1928), whose English edition, published in 1939 under the title *Probability, Statistics and Truth*, became very influential among scientists and probabilists. He outlined his own philosophical position in the volume *Kleines Lehrbuch der Positivismus* (1939), published in English as *Positivism* (1951), also containing a chapter on probability. A number of more technical contributions were collected by his wife Hilda Geiringer in the volume *Mathematical Theory of Probability and Statistics*, published posthumously

---

[5] An extensive and competent account of logical empiricism is to be found in Stadler (1997).

[6] Reported in Stadler (1997, English edition 2001), p. 689.

in 1964.

In the first chapter of *Probability, Statistics and Truth* von Mises writes

> We state here explicitly: The rational concept of probability, which is the only basis of probability calculus, applies only to problems in which either the same event repeats itself again and again, or a great number of uniform elements are involved at the same time. Using the language of physics, we may say that in order to apply the theory of probability we must have a practically unlimited sequence of uniform observations (von Mises 1928, English edition 1939, 1957, p. 11).

As suggested by this passage, von Mises is not interested in the intuitive notion of probability occurring in everyday speech, but in a 'rational' concept of probability, susceptible of an unambiguous definition and devised for a well defined field of application. He makes it clear at the outset that the only notion of probability that can be deemed 'rational' in the sense described, is in terms of frequency, calculated with reference to mass phenomena, namely to phenomena resulting from a great number of elements, or consisting of indefinitely repeatable events. Suchlike phenomena are encountered in 'games of chance', in 'certain problems relating to social mass phenomena' and, last but not least, in 'certain mechanical and physical phenomena' (*ibid*, p. 10).

As an immediate consequence of this approach, to speak of probability with reference to a single event, like the death of a person, the throw of a coin, the behaviour of a gas molecule, makes sense only if one refers to the class to which the individuals in question belong. To stress this feature of the frequency approach, von Mises says that talking of the probability of single events 'has no meaning' (*ibid.*, p. 11). This expression echoes the terminology adopted by logical empiricists, who labelled 'meaningless' the concepts belonging to traditional metaphysics.

## Collectives

To refer to mass phenomena, von Mises adopts the term *collective*, taken to denote 'a sequence of uniform events or processes which differ by certain observable attributes, say colours, numbers, or anything else' (*ibid.*, p. 12). To exemplify the notion of collective, he adds that

> All the peas grown by a botanist concerned with the problem of heredity may be considered as a collective, the attributes in which we are interested being the different colours of the flowers. All the throws

of dice made in the course of a game form a collective wherein the attribute of the single event is the number of points thrown. Again, all the molecules in a given volume of a gas may be considered as a collective, and the attribute of a single molecule might be its velocity. A further example of a collective is the whole class of insured men and women whose ages at death have been registered by an insurance office (*ibidem*).

The notion of collective assumes a prime importance within von Mises' theory of probability, which becomes a theory of collectives in the sense that collectives represent the object of probability theory as well as its scope of application. In the author's words:

> The principle which underlies the whole of our treatment of the probability problem is that a collective must exist before we begin to speak of probability. The definition of probability which we shall give is only concerned with 'the probability of encountering a certain attribute in a given collective' (*ibidem*).

After having introduced the collective by way of examples, as just described, von Mises moves to a more precise definition, involving the notion of probability as relative frequency. According to this definition,

> a collective is a mass phenomenon or a repetitive event, or, simply, a long sequence of observations for which there are sufficient reasons to believe that the relative frequency of the observed attribute would tend to a fixed limit if the observations were indefinitely continued. This limit will be called *the probability of the attribute considered within the given collective* (*ibid.*, p. 15).

Von Mises stresses that the existence of a collective is a necessary condition for probability, in the sense that without a collective there cannot be any meaningful probability assignments. This is reflected by the title of a section of the first chapter of *Probability, Statistics and Truth*: 'First the collective – then the probability' (*ibid.*, p. 18). Von Mises considers this the only non ambiguous way of addressing probability, and regards all other approaches as ill-founded[7].

This definition of probability holds only provided that the relative

---

[7] In this connection, von Mises criticizes Johannes von Kries for taking 'the diametrically opposite viewpoint' that one can genuinely talk of the probability of single occurrences of events. See von Mises (1928, English edition 1939, 1957) p. 18.

frequencies of observed attributes have a limit. Von Mises is aware that this is a strong assumption, but believes it to be fully justified in a great many cases, both in science and everyday life. In the case of games of chance, for example, 'the hypothesis of the existence of limiting values of the relative frequencies is well corroborated by a large mass of experience' (*ibid.*, p. 16).

However, the limit assumption is by no means sufficient to characterize collectives satisfactorily. Examples can readily be found of sequences which exhibit relative frequencies tending to fixed limits, but nevertheless do not qualify as probabilistic. To instantiate this, von Mises invites us to

> Imagine [...] a road along which milestones are placed, large ones for whole miles and smaller ones for tenths of a mile. If we walk long enough along this road, calculating the relative frequencies of large stones, the value found in this way will lie around 1/10. The value will be exactly 0.1 whenever in each mile we are in that interval between two small milestones which corresponds to the one in which we started. The deviations from the value 0.1 will become smaller and smaller as the number of stones passed increases; that is, the relative frequency tends towards the limiting value 0.1. This result may induce us to speak of a certain 'probability of encountering a large stone' (*ibid.*, p. 23).

At closer inspection, however, the case under consideration is essentially different from probabilistic sequences like, say, the throws of a coin. Unlike the latter, the milestones sequence is predictable and

> obeys an easily recognizable law. Exactly every tenth observation leads to the attribute 'large', all others to the attribute 'small'. After having just passed a large stone, we are in no doubt about the size of the next one; there is no chance of its being large (*ibidem*).

The example brings us to the second basic requirement of collectives, namely 'lawlessness', or *randomness*. While regarding randomness as an essential feature of probabilistic sequences, von Mises reaffirms the theoretical priority of this notion over that of probability.

## Randomness

In order to produce a rigorous definition of randomness, von Mises adopts an operative notion named *place selection*. Let us take a sequence satisfying the first requirement imposed on collectives. This will be an infinite sequence of elements exhibiting a certain attribute, whose relative frequency, determined on the basis of the observation of $n$ elements, is $m/n$. The probability of such an attribute in the sequence will be the limit to which the relative frequency $m/n$ tends when $n \to \infty$. Now, let us imagine deriving from the given sequence an infinite sub-sequence. Any such sub-sequence can be said to be derived by place selection if the decision whether the $p^{th}$ term of the original sequence is to be included in the sub-sequence depends on the number $p$ and on the attributes of the $(p - 1)$ preceding elements, not on the attribute of the $p^{th}$ term or any other attribute following in the sequence. Each place selection is defined by a rule, stating unambiguously for every element of the sequence whether it ought to be made part of the sub-sequence or not. Examples of sub-sequences obtained by place selection are

> the partial sequences formed by all odd numbers of the original sequence, or by all members for which the place number in the sequence is the square of an integer, or a prime number, or a number selected according to some other rule, whatever it may be (*ibid.*, p. 25).

On the basis of this notion, von Mises defines randomness as *insensitivity to place selection*. This obtains when the limiting values of the relative frequencies in a given sequence are not affected by any of all the possible selections that can be performed on it and, in addition, the limiting values of the relative frequencies, in the sub-sequences obtained by place selection, equal those of the original sequence. As von Mises puts it, 'The limiting values of the relative frequencies in a collective must be independent of all possible place selections' (*ibidem*). This randomness condition is also called by von Mises the *principle of the impossibility of a gambling system*, because it reflects the impossibility of devising a system leading to a gain, in any hypothetical game. To von Mises, the failure of all the attempts aimed at devising gambling systems provides the empirical foundation of the notion of a random collective:

> Everybody who has been to Monte Carlo, or who has read descriptions of a gambling bank, knows how many 'absolutely safe' gambling systems, sometimes of an enormously complicated character, have been invented and tried out by gamblers; and new systems are still being suggested every day. The authors of such systems have all, sooner or later, had the sad experience of finding out that no system is able to improve their chances of winning in the long run, i.e., to affect the relative frequencies with which different colours or numbers appear in a sequence selected from the total sequence of the game. This experience

forms the experimental basis of our definition of probability (*ibidem*).

Therefore, the principle of randomness constitutes the empirical and operational basis of the notion of collective, and consequently of von Mises' theory of probability, grounded on it.

Von Mises' definition of randomness rigorously restates the long-standing idea that events are random when they are unpredictable and cannot be accounted for in causal terms. It should not pass unnoticed that while defining randomness in terms of insensitivity to *all* possible place selections, von Mises embraces an absolute, unrestricted notion of randomness. This choice is philosophically motivated by an urge to secure an objective foundation to probability. However, soon after it was proposed by von Mises, his theory of randomness raised serious objections, on which more will be said in the last section of Chapter 5. Anticipating what will follow, it can be added that the difficulties affecting von Mises' approach shed doubts on the whole project of defining randomness in absolute terms. In Section 4.4 we shall see how Hans Reichenbach held the view that von Mises' notion of randomness is exceedingly restrictive, and turned to a weaker notion.

## Collective-based probability

Having defined collectives, von Mises proceeds to spell out a notion of probability based on them. The first step in this direction consists in defining the notion of *distribution*, which is taken to denote 'the whole of the probabilities attached to the different attributes in a collective' (*ibid.*, p. 35). In other words, one can think of a distribution as the way in which

> the different possible attributes are distributed in the infinite sequence of elements forming the collective. If, for instance, the numbers 1/5, 3/5, and 1/5 represent the distribution in a collective with three attributes A, B, and C, the probabilities of A and C being 1/5 each, and that of B being 3/5, then in a sufficiently long sequence of observations we shall find the attributes A, B, and C 'distributed' in such a way that the first and third of them occur in 1/5 of all observed cases and the second in the remaining 3/5 (*ibidem*).

Along similar lines, von Mises defines the notion of continuous distribution. The notion of distribution has a fundamental role, because 'The purpose of the theory of probability is to calculate the distribution in the new collective from the known distribution (or distributions) in the initial ones' (*ibid.*, p. 37). The fundamental properties of probability are therefore defined in terms of operations that allow for the derivation of one collective from another. To this end, von Mises introduces four fundamental operations, called *selection*, *mixing*, *partition* and *combination*, and shows how by means of such operations the whole theory of probability can be re-stated in terms of collectives.

Von Mises regards his own theory as a decisive improvement over the classical definition of probability. He is especially critical of the notion of 'equally probable' or 'equally likely' cases, being convinced that outside of the restricted field of games of chance, when one deals for example with the probability of death, or of the throws of a biased die, there are no equally possible cases to be found. For instance,

> According to a certain assurance table [...] the probability that a man forty years old will die within the next year is 0.011. Where are the 'equally likely cases' in this example? Which are the favourable ones? (*ibid.*, p. 69).

In like situations, which are most likely to be the object of probability, one is compelled to resort to frequencies. Von Mises remarks that this is in fact what is usually done when, appealing to the law of large numbers, the frequency of some attribute in a sufficiently long series of experiments is calculated, 'passing, as if it were a matter of no importance, from the consideration of a priori probabilities to the discussion of cases where the probability is not known a priori' (*ibid.*, p. 70). This way of passing from the laws of probability defined in terms of equally possible cases, to probability assignments representing frequencies is deemed by von Mises ill-founded, and should be abandoned in favour of an approach explicitly grounded on frequencies. In his words:

> up to the present time, no one has succeeded in developing a complete theory of probability without, sooner or later, introducing probability by means of the relative frequencies in long sequences. There is, then, little reason to adhere to a definition which is too narrow for the inclusion of a number of important applications and which must be given a forced interpretation in order to be capable of dealing with many questions of which the theory of probability has to take cognizance (*ibidem*).

In developing his theory, von Mises' task is to provide a rigorous foundation to an empirical notion of probability, which can be operationally reduced to a measurable quantity. It can be objected to the operational character of von Mises' theory that it uses infinite sequences. This raises a problem of applicability, that is addressed both in *Probability, Statistics and Truth* and *Positivism*. As already noted, von Mises believes that probability as an idealized limit can be compared to other limiting notions to be encountered in science, such as velocity or density[8]. Having said that, he concedes that a problem of applicability may arise in connection with the relationship between the sequences of observations, which are obviously finite, and the infinite sequences postulated by the theory. Such a relationship involves a twofold passage: from observation to theory and vice versa. If the first passage leads to the formulation of theories, the second is made necessary by the need to compare the results with experience obtained through deductions in the theoretical axiomatic systems. The first passage involves an inductive step, which in the case of probability consists in recognizing the stability of relative frequencies and insensitivity to selection of subsequences. As to the second passage, von Mises claims that

> the results of a theory based on the notion of the infinite collective can be applied to infinite sequences of observations in a way which is not logically definable, but is nevertheless sufficiently exact in practice (*ibid.*, p. 85)[9].

## Applications to science

The last chapter of *Probability, Statistics and Truth* is entirely devoted to the application of von Mises' theory of probability to physical science. According to the author, probability can be applied to physical phenomena only insofar as they can be reduced to 'chance mechanisms' having the features of collectives. He mentions the following as typical areas of physics that can be treated probabilistically: 1. the kinetic theory of gases, 2. Brownian motion, 3. radioactivity, 4. Planck's theory of black-body radiation. He argues that in all these areas the frequency theory of probability, and more specifically his own version of it, applies naturally. After showing how the phenomena which are the object of the above theories can be reduced to collectives and treated

---

[8] See von Mises (1939, English edition 1951, 1968), p. 168.

[9] Oddly enough, von Mises associates a position of this kind with Carl Gustav Hempel, who seems rather committed to a finitistic version of frequentism. For Hempel's paper mentioned by von Mises see Hempel (1935, English edition 2000).

probabilistically, von Mises discusses 'the new quantum statistics created by de Broglie, Schrödinger, Heisenberg and Born' (*ibid.*, p. 211), and holds the view that frequentist probability can be extended to this field without major problems. As a matter of fact, the application of the frequency notion of probability to quantum theory appears problematic, to say the least. The main difficulty arises in connection with the problem of the single case, because within quantum theory it is common to talk about single case probabilities – for instance, the probability that a single atom is in a certain state – which is not contemplated by von Mises' theory.

Von Mises pushes his positivistic view of probability to its extreme consequences, by embracing indeterminism and denying the principle of causality. Regarding the latter, he claims that

> It now appears inevitable that we must abandon another cherished notion that has its origin in everyday life and pre-scientific thought and has been elevated to the rank of an eternal category of thought by overly zealous philosophers: the naive concept of causality (*ibid.*, p. 210).

Philosophers have given so many, and so vague, different formulations of the principle of causality, that 'it is not at all easy, perhaps hardly possible, to contradict this "law"' (*ibidem*). According to von Mises, this makes the principle itself useless. One is led to the same conclusion by considering the developments made by physical science, where recourse to statistical methods has gradually superseded causal talk. Von Mises' conclusion is that

> At such time when physics, and more generally natural science based on observations, shall have completely assimilated the methods and arguments of statistical theory and shall have recognized them as essential tools, the feeling will disappear that these methods and theories contradict any logical need, any 'necessity of thought', or that they leave some philosophical requirement unfulfilled. In other words, the principle of causality is subject to change and it will adjust itself to the requirements of physics (*ibid.*, p. 211).

The withdrawal of both causality and the deterministic world view traditionally associated with it is urged by Heisenberg's principle of uncertainty, which von Mises welcomes as a possible basis on which the old and the new physics could be unified on probabilistic grounds. After the advent of Boltzmann's statistical mechanics in the second half of the nineteenth century, it was in fact recognized that 'the predictions of classical physics are to be understood in the sense of probability statements of the type of the Laws of Large Numbers' (*ibid.*, pp. 217-8), but 'at this stage of development, the usual assumption was that the atomic processes themselves, namely the motions of single molecules, are governed by the exact laws of deterministic mechanics' (*ibid.*, p. 218). This dualism came to an end with quantum mechanics: 'the rise of quantum mechanics has freed us from this dualism which prevented a logically satisfactory formulation of the fundamentals of physics' (*ibidem*). After these developments, the whole edifice of science can be made to rest on a statistical conception of nature, granting indeterminism the same credibility traditionally ascribed to determinism.

The book *Probability, Statistics and Truth* ends with a passage, in which von Mises epitomizes his thesis, reflected by the words of its title:

> Starting from a logically clear concept of *probability*, based on experience, using arguments which are usually called *statistical*, we can discover *truth* in wide domains of human interest (*ibid.*, p. 220).

Von Mises' message is that statistical methods, rooted in the frequency notion of probability, provide a powerful heuristic tool for the investigation of reality. This conviction inspires a probabilistic approach to the construction of scientific knowledge, in which the desire for objectivity goes hand in hand with an ingrained empiricism. Frequentism received great impulse from the work of von Mises, who was well known and widely respected within scientific circles. After von Mises, the frequentist interpretation of probability, already quite popular among statisticians, geneticists and researchers in the social sciences, also became popular among physicists, who often refer to his works in their writings.

## 4.4 Hans Reichenbach's probabilistic epistemology

### Reichenbach's frequentism

A slightly modified version of frequentism was developed by Hans Reichenbach (1891-1953), another logical empiricist and member of the 'Berlin Society for empirical philosophy'. Born in Hamburg, Reichenbach studied engineering, mathematics, physics and philosophy at various universities, and in 1926 became professor of philosophy and physics in Berlin.

In 1933 he was dismissed from his job for racial reasons, and – like von Mises – emigrated to Istanbul. He remained there until 1938, when he finally moved to Los Angeles. There he held a professorship at the University of California, until he died of a heart attack in 1953. Reichenbach was one of the most active supporters of logical empiricism, and in 1930 became, together with Rudolf Carnap, editor of *Erkenntnis*, the journal of the movement. In a portrait, Wesley Salmon depicts Reichenbach as an extraordinarily gifted man, 'a staunch supporter of freedom of enquiry and education' and 'a passionate advocate of individual self-determination of moral goals as well as intellectual deliberation' (Salmon 1979a, p. 5).

Reichenbach produced many articles and books on various subjects belonging to philosophy of science, and in particular to philosophy of physics. His writings on probability include the books *Wahrscheinlichkeitslehre* (1935), later merged into *The Theory of Probability* (1949), and *Experience and Prediction* (1938). Probability was one of Reichenbach's chief interests, that deeply imbued his epistemology. Straying from the mainstream tendency of logical empiricism, Reichenbach takes a sceptical attitude towards the notion of truth, in which connection he claims that

> The ideal of an absolute truth is [...] a phantom, unrealizable; certainty is a privilege pertaining only to tautologies, namely those propositions which do not convey any knowledge (Reichenbach 1937, p. 90).

For Reichenbach it is probability, rather than truth, that can substantiate a viable theory of knowledge, and a reconstruction of science in tune with scientific practice.

While embracing a frequency theory of probability, Reichenbach emphasizes that his position should not be conflated with that of von Mises, or regarded as a continuation of the latter. In a letter to Bertrand Russell written in 1949, referring to Russell's book *Human Knowledge*, Reichenbach writes:

> I was surprised to find myself hyphenated to von Mises [...] – as much surprised, presumably, as he. You even call my theory a development of that of von Mises. I do not think this is a correct statement. My first publication on probability [*Der Begriff der Wahrscheinlichkeit für die mathematische Dartstellung der Wirklichkeit*, Leipzig, 1915], which is earlier than Mises' publications, has already a frequency interpretation and a criticism of the principle of indifference, although later I abandoned the Kantian frame of this paper. [...] Mises' merit is to have shown that the strict-limit interpretation does not lead to contradictions

and, further, to have provided a means for the characterization of random sequences. I then could show that my earlier frequency interpretation (which was weaker than a strict-limit interpretation) in combination with Bernoulli's theorem leads to the limit interpretation and thus took over this interpretation. But my mathematical theory is more comprehensive than Mises' theory, since it is not restricted to random sequences; furthermore, Mises does not connect his theory with the logical symbolism. And Mises has never had a theory of induction or of application of his theory to physical reality (Reichenbach 1978, vol. 2, p. 410).

Indeed, Reichenbach's frequentism is more flexible than that of von Mises, because it allows for single case probabilities, develops a theory of induction and contains an argument for its justification.

A first difference between von Mises' and Reichenbach's perspectives lies in the notion of randomness. Reichenbach, who was well aware of the difficulties faced by von Mises' definition, was less concerned with the mathematical definition of randomness, than with its practical applications. He then turned to a weaker concept of randomness, which is not based on an absolute invariance domain, but is relative to a restricted domain of selections 'not defined by mathematical rules, but by reference to physical (or psychological) occurrences' (Reichenbach 1935, English edition 1949, 1971, p. 150). According to this approach,

Random sequences are characterized by the peculiarity that a person who does not know the attributes of the elements is unable to construct a mathematical selection by which he would, on an average, select more hits that would correspond to the frequency of the major sequence. In other words, such selections will be included in the domain of invariance. In this form, the impossibility of making a deviating selection is expressed by a psychological, not a logical, statement; it refers to acts performed by a human being. This may be called a *psychological randomness* (*ibidem*).

Reichenbach is aware of the weakness of a definition that 'instead of speaking of logical impossibilities, refers only to a limitation of the technical abilities of human observers', but adds that 'such a psychological reference is indispensable, too, when selections in terms of physical observations are to be incorporated in the domain of invariance' (*ibidem*).

With an eye to scientific practice, Reichenbach observes that 'the significance of the problem of the definition of random sequences should not be overestimated', because 'random sequences represent merely a special type' of probability sequences which are tractable by means of statistical methodology. His conclusion is that

> All types of probability sequences are found in nature. A mathematical theory of probability should not be restricted to the study of one specific type of sequence but should include suitable definitions of various types, chosen from the standpoint of practical use (*ibid.*, p. 151).

Also for practical purposes, Reichenbach introduces the notion of *practical limit* 'for sequences that, in dimensions accessible to human observation, converge sufficiently and remain within the interval of convergence', and adds that 'it is with sequences having a practical limit that all actual statistics are concerned' (*ibid.*, pp. 347-8).

Like von Mises, Reichenbach embraces an empiricist view of probability, according to which degrees of probability can never be ascertained *a priori*, but only *a posteriori*. The method by which degrees of probability are obtained is *induction by enumeration*. This

> is based on counting the relative frequency [of a certain attribute] in an initial section of the sequence, and consists in the inference that the relative frequency observed will persist approximately for the rest of the sequence; or, in other words, that the observed value represents, within certain limits of exactness, the value of the limit for the whole sequence (*ibid.*, p. 351).

The procedure described amounts to *inductive inference*, to be stated in more precise terms by means of the *rule of induction*:

> If the sequence has a limit of the frequency, there must exist an $n$ such that from there on the frequency $f^i$ ($i > n$) will remain within the interval $f^n \pm \delta$, where $\delta$ is a quantity that we can choose as small as we like, but that, once chosen, is kept constant. Now if we posit that the frequency $f^i$ will remain within the interval $f^n \pm \delta$, and we correct this posit for greater $n$ by the same rule, we must finally come to the correct result (*ibid.*, p. 445).

According to Reichenbach's definition, probability is inextricably intertwined with induction. His formulation of the rule of induction is meant precisely to make it explicit that an inductive step is required, in order to define probability as limiting frequency. As we know from Reichenbach's letter to Russell, he regarded it as a definite improvement over von Mises that a theory of induction is embodied in his own theory of probability.

## The theory of posits

As suggested by the rule of induction, for Reichenbach any probability attribution is a *posit*, namely 'a statement with which we deal as true, although the truth value is unknown' (*ibid.*, p. 373). The notion of posit occupies a crucial role within Reichenbach's theory of probability, where it is introduced by analogy with gambling:

> The gambler has to make a prediction before every game, although he knows that the calculated probability has a meaning only for larger numbers; and he makes his decision by betting, or as we shall say, by *positing* the more probable event. [...] The frequency interpretation justifies, indeed, a *posit* on the more probable case. It is true that it cannot give us a guarantee that we shall be successful in the particular instance considered; but instead it supplies us with a principle which in repeated application leads to a greater number of successes than would obtain if we acted against it (*ibid.*, p. 314).

As suggested by this passage, the notion of 'success' plays an important role in Reichenbach's conceptual construction, to which it confers an aura of pragmatism.

An important feature of the notion of posit lies in the fact that it 'represents the bridge between the probability of the [...] sequence and the compulsion to make a decision in a single case' (*ibid.*, p. 315). Reichenbach tries to cope with the problem of the single case posed by von Mises' theory. The idea here is that a posit regarding a single occurrence of an event receives a *weight* from the probabilities attached to the reference class to which the event in question has been assigned, which must obey a criterion of homogeneity. That is to say that the reference class should be chosen in such a way as to include as many as cases as possible similar to the one under consideration, while excluding dissimilar ones. Similarity is to be taken as relative to the properties that are considered to be relevant, and homogeneity is obtained through successive partitions of the reference class, by means of statistically relevant properties. Once a reference class cannot be further partitioned in this way, it is said to be *homogeneous*. For instance, if one wanted to assign a weight to the probability that a given individual – say William White – will die within the next five years, the reference class ought to be chosen on the basis of a series of properties taken to be relevant, like age, sex, occupation, nationality, health status, and so on, whereas irrelevant properties, like hair colour, or shoe size, should be excluded. The probability of death of individuals belonging to the reference class so obtained, determinable on the basis of observed frequencies, will give the weight to be assigned to the survival of the specific individual William White. '*A weight* – Reichenbach says – *is what a degree of probability becomes if it is applied to a single case*' (Reichenbach 1938, p. 314).

An attempt to accommodate within frequentism single case probability attributions is certainly a merit of Reichenbach's theory. However, the proposed solution is not free from difficulties, because, in principle, one can never be absolutely sure that all the properties that are relevant to some phenomenon have been taken into account. Identifying the proper reference class is obviously a delicate matter, which poses serious problems, well-known and widely discussed by statisticians.

Reichenbach's posits differ depending on whether they are made in a situation of *primitive* or *advanced* knowledge. When some knowledge of probabilities is available, the state of knowledge is called 'advanced', otherwise it is 'primitive'. Within primitive knowledge, use of the rule of induction yields probability values, while 'all the questions concerning induction in advanced knowledge [...] are answered in the calculus of probabilities' (Reichenbach 1935, English edition 1949, 1971, p. 432). Posits made in a state of advanced knowledge have a definite weight, and are called *appraised*. They conform to the principle of the greatest number of successes, which makes them the best posits that can be made. A posit whose weight is unknown is called *anticipative*, or *blind*. Although the weight of a blind posit is unknown, its value can be corrected. Blind posits have an approximate character: we know that by making and correcting such posits we will eventually achieve success, if the sequence has a limit. Reichenbach grounds on this idea his argument for the justification of induction, to be outlined later. Scientific hypotheses are confirmed within the framework of advanced knowledge, and made to rely on the Bayesian method. The pivotal role assigned in this connection to Bayes' rule is a distinctive feature of Reichenbach's approach, according to which prior probabilities should be determined on the basis of frequencies. In view of this, Reichenbach qualifies as an 'objective Bayesian'.

Reichenbach regarded the whole edifice of knowledge as genuinely probabilistic, resulting from a continuous interplay between blind and appraised posits. In this framework, posits form a hierarchical system, in which they stand on different levels:

> if we know the limit of the frequency in a sequence, this value can be regarded as the weight of an individual posit concerning an unknown element of the sequence. The weight may be identified with the probability of the single case, assuming the character of a truth value. In order to find the limit of the frequency we use an anticipative posit; its weight is unknown. In order to determine this weight we must make an anticipative posit on a higher linguistic level; the former anticipative posit is then transformed into an appraised posit, that is, a posit of known weight. The procedure can be extended to higher and higher levels (*ibid.*, p. 465).

Scientific knowledge so conceived retains a peculiar self-correcting character, derived from the analogous character of the rule of induction:

> By means of the inductive rule we set up posits concerning the limit of the frequency in a sequence and thus establish probability values. The probabilities so constructed can be used as the weights of certain other posits; we are thus able to construct appraised posits by means of anticipative posits. The appraised posits can even be identical with some of the anticipative posits; in other words, we can transform an anticipative posit into an appraised posit. Since the weight thus constructed can be used for a change in the posited value of the limit, we speak here of the *method of correction* (*ibid.*, p. 461).

Knowledge acquisition rests on this kind of self-correcting procedure, which starts with blind posits, and goes on to formulate appraised posits that become part of a complex system. The latter can be taken as a representation of science itself: 'the system of science [...] must be regarded as a system of posits' (*ibid.*, p. 469). As echoed by the title of one of his major works, namely *Experience and Prediction*, for Reichenbach scientific knowledge results from the interplay between these two fundamental components. The method of posits, passing from experience of frequencies to predictions on probabilities, reflects such an interplay and provides the fundamental ingredient of scientific method. As Reichenbach puts it: 'scientific method is nothing but a continuous correction of posits by incorporating them into more general considerations' (Reichenbach 1933, English edition 1949, p. 318). The holistic character of Reichenbach's conception of scientific method should not pass unnoticed:

The system of scientific knowledge can be conceived as a method of correction, which relates every individual statement to the total system of experience. It is the significance of scientific method that for the prediction of a new phenomenon we are never dependent on the specific observations alone to which the prediction refers, but that we can also make use of the vast domain of experiences in very different fields (*ibidem*).

In one sentence, the edifice of science is deeply rooted in frequencies, and scientific method rests ultimately on the rule of induction.

## The justification of induction

This view of scientific knowledge calls for an argument for the justification of induction. To this end, Reichenbach adopts a pragmatic approach, making use of the notion of success. His argument is based on the approximate character of blind posits: we know that by making and correcting such posits we will eventually reach success, in case the considered sequence has a limit. Since the method of blind posits rests on the rule of induction, Reichenbach's argument applies to the latter, and says that

> the rule of induction is justified as an instrument of positing because it is a method of which we know that if it is possible to make statements about the future we shall find them by means of this method (1935, English edition 1949, 1971, p. 475).

Starting from the idea that induction cannot be justified logically, Reichenbach attempts to 'vindicate' it on pragmatic grounds, on the basis of the consideration that it is a necessary condition for making predictions.

To illustrate his viewpoint on the justification of induction, Reichenbach makes use of a metaphoric image, which ends the volume *The Theory of Probability*:

> A blind man who has lost his way in the mountains feels a trail with his stick. He does not know where the path will lead him, or whether it may take him so close to the edge of a precipice that he will be plunged into the abyss. Yet he follows the path, groping his way step by step; for if there is any possibility of getting out of the wilderness, it is by feeling his way along the path. As blind men we face the future; but we feel a path. And we know: if we can find a way through the future it is by feeling our way along this path (*ibid.*, p. 482).

Reichenbach's solution to the problem of induction is in tune with the position assumed by another logical empiricist, namely Herbert Feigl. This author, who was member of the Vienna Circle, made a distinction between two kinds of justifying procedures, called 'vindication' and 'validation'. The validation procedure, commonly used within deductive logic, consists in appealing to more and more general standards, until the fundamental principles of a theory are reached. To justify such principles, what is needed is a validation by means of pragmatic considerations, typically based on success in view of the achievement of a certain goal. Given that the task of induction is to acquire new knowledge while formulating successful predictions, Feigl proposes to regard an inductive method as vindicated, if it can be shown that it allows for correct predictions about the future[10].

The use of pragmatic arguments for the justification of induction looks to many authors like the only possible way out of Hume's problem. Among them Wesley Salmon, who has made an attempt to improve Reichenbach's argument in various ways. Making use of some results obtained by Ian Hacking, Salmon tried to supply Reichenbach's argument, which in fact applies to a whole class of asymptotic rules, with further conditions, apt to restrict it to the sole rule of induction[11]. Despite the efforts made by these authors, no conclusive solution has so far been attained.

## Causality

Reichenbach's epistemology is deeply probabilistic. In his view, the logic of science is not the two-valued classical logic based on truth and falsehood, but a probabilistic logic: 'the alternatives in science are not truth and falsehood; instead, there is a continuous scale of probability values whose unattainable limits are truth and falsehood' (Reichenbach 1930, English edition 1959, 1978, vol. 2, p. 342). In this spirit, he also attaches a probabilistic meaning to the notion of causality. His pioneering ideas in this connection heralded a fruitful field of enquiry. The basic idea underpinning his notion of probabilistic causality is that of defining the notion of causal relevance on the basis of statistical relevance, with some restrictions devised to entertain the fact that causal relations are asymmetrical, while statistical relevance relations are symmetrical. Reichenbach grounds the asymmetry of time on the asymmetry of

---

[10] See Feigl (1950).

[11] For a survey of the literature on the topic, including an interesting proposal, see Salmon (1991).

causal relations, and develops a causal theory of time in which the direction of time, as well as the notion of temporal priority, are defined on the basis of causal asymmetry and antecedence[12].

The fundamental principle of Reichenbach's causal theory of time is the *principle of the common cause*. In brief, it says that if two (or more) events of a certain kind happen jointly – though in different places – more often than would be expected were they independent, then this apparent coincidence should be explained in terms of a common causal antecedent. This principle has been taken up by Wesley Salmon, and put at the core of a theory of scientific explanation, taking inspiration from Reichenbach's idea that 'the causal structure of the universe can be comprehended with the help of the concept of *probable determination* alone' (Reichenbach 1925, English edition 1978, vol. 2, p. 83)[13].

Within Reichenbach's probabilistic epistemology, the traditional distinction between probabilistic regularities and causal laws is 'only superficial', because 'probability laws and causal laws are logical variations of one and the same type of regularity' (Reichenbach 1930, English edition 1959, 1978, vol. 2, p. 335). General statements about natural phenomena are retained subjected to various conditions, and involve some degree of uncertainty. As Reichenbach puts it:

> The characterization of the causal laws of nature as strict laws is justified only for certain schematizations. When all causal factors are known, then an effect can be predicted with certainty; such an idealization would be irrelevant for science, without the addition of further assumptions. It is impossible to know all causal factors; we can only select a limited number of relevant factors and use them to predict future events, but must neglect factors of lesser influence. It is usually assumed that the influence of the less important factors is small, and that we can therefore predict the future within certain limits of exactness. This formulation is inadequate, however, and misses a fundamental point in the epistemological situation. Actually, we can only maintain that it is highly probable that future events will lie within certain limits of exactness (*ibidem*).

---

[12] For Reichenbach's causal theory of time see his (1956).
[13] Salmon's theory of explanation falls out of the scope of this book. See Salmon (1971), (1984) and (1998) for further reading.

Reichenbach's stress on prediction is a distinctive feature of his perspective, where the theory of probability is put forward as a 'theory of propositions about the future' (Reichenbach 1936, p. 159).

To conclude this presentation of Reichenbach's perspective, it should be observed that, though more flexible than that of von Mises, Reichenbach's version of frequentism has not encountered much success among statisticians and scientists, who have by and large ignored it. On the other hand, Reichenbach's theory of probability is philosophically very insightful and suggestive of new developments, as testified, among other things, by Salmon's work on explanation.

## 4.5 Ernest Nagel's truth-frequency theory

Ernest Nagel (1901-1985) is one of the most prominent philosophers of science of the last century. Therefore, a brief recollection of his views on probability does not seem out of place. Born in Bohemia, Nagel emigrated with his family to the United States at the age of ten. He obtained a doctorate in philosophy from Columbia, where he later held a professorship for many decades, and continued to teach after retirement[14]. Nagel's philosophical perspective combines elements of pragmatism – especially in John Dewey's version – with other ingredients typical of logical empiricism, such as analytic method and formal logic.

Nagel wrote the volume on probability for the 'International Encyclopedia of Unified Science', conceived by Otto Neurath as an instrument of integration among the various fields of enquiry, in the spirit of that 'scientific conception of the world' fostered by the logical empiricists. In that work, published in 1939 under the title *Principles of the Theory of Probability*[15], Nagel takes sides with the frequentist theory of probability, but at the same time he makes 'a sustained effort to survey the various alternatives [to frequentism] and to deal with many of the problems that have been raised about the frequency interpretation' (Suppes 1999, p. 217). In this spirit, Nagel refers to a number of upholders of alternative views, like Keynes, Ramsey, and even de Finetti, who at the time was not yet very renowned.

Nagel embraces a peculiar version of frequentism, which strays in various

---

[14] For a profile of Ernest Nagel written by one of his most famous pupils, see Suppes (1999).

[15] See Nagel (1939a).

respects from the perspective of both von Mises and Reichenbach. Under the influence of the doctrine of 'leading principles' formulated by Charles Sanders Peirce, Nagel embraces a *truth-frequency* theory of probability, according to which probability refers to an inference from one set of propositions to another, and denotes the relative frequency of the effectiveness of such an inference. After defining a leading principle as 'a proposition formulating a class of inferences' (Nagel 1933, p. 537), Nagel maintains that

> the phrase 'the probability that heads will come up' has meaning only in terms of the *type* of argument employed to infer the proposition 'heads will come up'. If that type of inference deserves our confidence, it is because in fact it does lead in a certain proportion of cases to true conclusion from true premises (*ibid.*, p. 538).

This leads to a conception of probability as 'a *relation* between *propositions*', whose meaning is made to reside in frequencies:

> It is a relation whose analysis shows that the relative truth-frequencies of classes of propositions between which it holds *is* part of its meaning. It follows, therefore, that a proposition is correctly judged to have a probability only in so far as it is referred to some *class* of propositions, and that a proposition will be judged to have different degrees of probability according as it is referred to one such class rather than another (*ibid.*, pp. 538-9).

In Nagel's perspective probability is taken as a theoretical notion, and probability statements are tested by comparing their consequences with observed frequencies. This brings Nagel's interpretation of probability close to the 'propensity' theory of probability, which – as we shall see in the next chapter – was anticipated by Peirce and later revived by Popper. The analogy with Popper's view comes also in connection with Nagel's idea that probability hints at physical situations. In Nagel's words:

> the 'truth frequency' theory of probability can not be only a logical theory. If its interpretation of the nature of probability is sound, the effectiveness of probable arguments should be grounded in certain generic traits of existence. The theory has metaphysical bearings as well as a logical function. It points to the existence of a certain type of natural structure and contributes toward the understanding of the objective nature of universals (*ibid.*, p. 551).

This attitude is somehow at odds with Nagel's pragmatic stance, according to which 'the term "probability" is not univocal, for it has different meanings in different contexts' (Nagel 1936a, p. 26). The unifying character of the different uses of probability in different contexts, derives from the fact that it represents a measure of success of a certain type of inference. But the classes of propositions to which such an inference applies are determined in ways that vary according to the context in which they occur. Similar considerations hold for confirmation, which represents one of the 'contexts' to which probability applies.

The context-dependence of the notion of probability answers the need, deeply felt by Nagel, to safeguard its applicability. It is precisely for that reason that he regards Reichenbach's version of frequentism as having 'obvious methodological advantages over the theory of von Mises' (Nagel 1936b, p. 503). So much granted, he moves various objections to Reichenbach's theory. For instance, he objects to Reichenbach's notion of weight that 'the weight of a proposition cannot be definitely established by any finite number of observations' (Nagel 1939b, p. 217). Furthermore, against both Reichenbach and von Mises, Nagel maintains that probabilities are not obtained only through the observation of sequences, but can also be obtained otherwise; for instance, they can be deduced from established theories.

Where, according to Nagel, the truth-functional frequentist view marks a decisive advantage over Reichenbach's and Carnap's views, is in connection with the confirmation of general hypotheses. Reichenbach's frequentism is regarded as completely missing the point of what scientists mean when they speak of the evaluation of theories. 'I do not persuade myself – Nagel says – that Professor Reichenbach's interpretation of "the probability of a theory" formulates what physicists do with it or even think about it for the most part' (*ibid.*, p. 230). The crucial point overlooked by Reichenbach, lies with the fact that, when evaluating hypotheses, scientists make use of 'numerical measures *obtained under specified conditions of experiment*' (*ibidem*). By contrast, Nagel's truth-frequency interpretation is defined with explicit reference to the experimental context, and this is a further trait it shares with Popper's propensity theory. Carnap's inductive logic – to be dealt with in Chapter 6 – is also criticized by Nagel for being distant from scientific practice, and because of the difficulties it faces in connection with the confirmation of general hypotheses. However, Nagel's arguments in this connection are too detailed to be given at this juncture[16].

As stated already, Nagel's theory of probability has not received much attention, neither from scientists, nor from philosophers of science, though it undeniably harbours some interesting and original traits.

---

[16] For Nagel's criticism of Carnap see Nagel's article and Carnap's 'Replies' in Schilpp, ed. (1963).

# 5

## *The propensity interpretation*

### 5.1 Peirce, the forerunner

The American philosopher Charles Sanders Peirce (1839-1914), is widely considered to have anticipated what is nowadays called the 'propensity' interpretation of probability. Son of the well known mathematician Benjamin, Peirce was educated at Harvard, where he graduated in chemistry. Appointed lecturer first at Harvard and later at Johns Hopkins University, around the mid-1880s he lost his job. Thereafter, he embarked on a life of retreat in a rural surroundings, occasionally interrupted by lectures, mainly given at Cambridge Massachusetts, on invitation of his close friend William James. Peirce and James can be considered the major representatives of American pragmatism. Peirce left a bulk of papers, various collections of which have appeared in print.

A multifarious personality, Peirce had innovative ideas on many topics, and greatly influenced a number of last century's thinkers. As Theodore Porter observed, he 'wished to make statistical method central to scientific reasoning' (Porter 1986, p. 221) and argued against determinism[1]. As to the interpretation of probability, Peirce claims that 'probability never properly refers immediately to a single event, but exclusively to the happening of a given kind of event on any occasion of a given kind' (Peirce 1910, 1932, p. 404). However, probability does not simply refer to past occurrences, and cannot therefore be calculated simply by taking the ratio between the number of occurrences of the event and the number of observed cases. Probability is rather 'the ratio that

---

[1] See Porter (1986), especially Chapter 7, where Peirce's argument against determinism is given in some detail, and the last chapter of Hacking (1990).

there *would be* in the long run' (*ibid.*, p. 405). To exemplify his idea of the matter, Peirce takes the statement that

> the *probability* that if a die be thrown from a dice box it will turn up a number divisible by three, is one-third. The statement means that the die has a certain 'would-be'; and to say that a die has a 'would-be' is to say that it has a property, quite analogous to any *habit* that a man might have. Only the 'would-be' of a die is presumably as much simpler and more definite than the man's habit as the die's homogeneous composition and cubical shape is simpler than the nature of the man's nervous system and soul; and just as it would be necessary, in order to define a man's habit, to describe how it would lead him to behave and upon what sort of occasion – albeit this statement would by no means imply that the habit *consists* in that action – so to define the die's 'would-be', it is necessary to say how it would lead the die to behave on an occasion that would bring out the full consequence of the 'would-be'; and this statement will not of itself imply that the 'would-be' of the die *consists* in such behaviour (*ibid.*, pp. 409-10).

It is because of the stress he puts on the *would-be*, or on other words describing the dispositional character of probability, that Peirce can be seen as anticipating the propensity theory. So much granted, the limit of Peirce's view lies with the fact that he regards the dispositional property of probability as pertaining to objects – such as the die of the above example[2] – while a fully-fledged propensity interpretation ascribes probability to the set of conditions surrounding the occurrence of events. This crucial step was taken by Popper, who is regarded as the first to embrace this interpretation.


## 5.2 Popper's propensity interpretation

### Falsificationism

Karl Raimund Popper (1902-1994) is one of the major protagonists of twentieth century philosophy. Born in Vienna, where he was educated in many disciplines, including mathematics, physics, psychology, philosophy and music, Popper emigrated to New Zealand to escape Nazism in 1937 and taught

---

[2] This is observed by Gillies (2000), p. 118.

for ten years at Canterbury University College in Christchurch. In 1947 he moved to London, where he was first reader, then professor at the London School of Economics until he retired in 1969. He received many honours, including election to the Royal Society – a unique privilege for a twentieth century philosopher[3] – and was very active until his death, writing on a wide range of topics, spanning philosophy, psychology, quantum mechanics and politics.

From the beginning of his career in philosophy of science, Popper opposed the inductivism of thinkers of the Vienna Circle, defending in contrast a deductivist approach to scientific knowledge, inspired by the conviction that theories are corroborated through a process of attempted, but failed falsification. Popper's *falsificationism* was spelled out in his *Logik der Forschung* (1934), later published in English under the title *The Logic of Scientific Discovery* (1959), and in a series of essays, especially those published in the collection *Conjectures and Refutations* (1962). In his intellectual autobiography, Popper claims to have reached this conclusion very early[4]:

> My main idea in 1919 was this. If somebody proposed a scientific theory he should answer, as Einstein did, the question: 'Under what conditions would I admit that my theory is untenable?' In other words, what conceivable facts would I accept as refutations, or falsifications, of my theory? (Popper 1974, p. 32).

He held this conviction throughout his life, developing a scientific methodology aimed at the falsification of scientific hypotheses, rather than their confirmation on an inductive basis. According to this perspective, a hypothesis is stated by way of a conjecture, and some way of refuting it is sought through a comparison with experimental evidence. In case observational data lead to a denial of one of the consequences of such hypothesis, the hypothesis itself is falsified. According to Popper, the decisive advantage of

---

[3] David Miller points out that Popper was in fact the only scholar of the last century to be elected member of the Royal Society for his contribution to philosophy. All the others, including Alfred North Whitehead and Bertrand Russell, were elected members for their contributions to mathematics and logic, not to philosophy. See Miller (1997), containing a short biography and a philosophical portrait of Karl Popper.

[4] Popper's claim to this effect is rebutted in Hacohen (2000), and ter Hark (2002) and (2004). Both authors argue that Popper's anti-inductivism originated later than he claims in his autobiography. The work of ter Hark was pointed out to me by Donald Gillies.

falsification over inductive confirmation is that it is a conclusive method, because one negative instance is enough to discard the initial hypothesis as false, while it would require an infinity of positive instances to establish inductively the truth of a general hypothesis. A pivotal role is ascribed to the testability of hypotheses, which is a function of the 'empirical content' of statements. The latter is defined as the set of the statements forbidden by a hypothesis, or, equivalently, as 'the class of its potential falsifiers'. Given two theories,

> if the class of potential falsifiers of one theory is 'larger' than that of the other, there will be more opportunities for the first theory to be refuted by experience; thus compared with the second theory, the first theory may be said to be 'falsifiable in a higher degree'. This also means that the first theory *says more* about the world of experience than the second theory, for it rules out a larger class of basic statements (Popper 1934, English edition 1959, 1968, pp. 112-3).

Popper's falsificationist methodology of 'conjectures and refutations' includes a notion of corroboration, defined in terms of resistance to severe tests. As Popper puts it, the idea underpinning the notion of *degree of corroboration* is 'to sum up, in a short formula, a *report* of the manner in which a theory has passed – or not passed – its tests, including an evaluation of the severity of the tests' (Popper 1974, p. 82). Degree of corroboration increases with the severity of tests, therefore only theories with a high degree of testability can be highly corroborated. Popper was always careful to keep corroboration separate from probability. Since hypotheses with a higher content are harder to confirm, degree of corroboration is actually 'linked to *improbability* rather than to *probability*' (*ibidem*). Albeit degree of corroboration, which can vary between −1 and +1, is not a probability, it is a function of probabilities. As a matter of fact, the probabilities that enter into the determination of degrees of corroboration are the same as those used by Bayesians, namely the probabilities of a certain hypothesis and a given piece of evidence, and the likelihood of the evidence relative to the hypothesis in question. On this basis, a number of authors have compared Popper's corroboration method with Bayesian confirmation, in spite of Popper's fierce opposition to Bayesianism[5].

---

[5] See, among others, Gillies (1998), Festa (1999), and Kuipers (2000).

## The propensity interpretation of probability

In two papers of the late Fifties, namely 'The Propensity Interpretation of the Calculus of Probability, and the Quantum Theory' (1957) and 'The propensity Interpretation of Probability' (1959), Popper proposed the propensity interpretation of probability, in an attempt to answer the problem of interpreting quantum mechanical probabilities. Later on, he no longer intended the notion of propensity only for application to quantum mechanics, making it the focus of a wider programme, meant to account for all sorts of causal tendencies operating in the world. Popper deems propensity 'purely objective', in view of the fact that propensities are taken as 'physically real'. Furthermore, propensities are 'metaphysical', because they refer to non-observable properties.

In the Fifties, the propensity interpretation was presented by Popper as a variant of the frequency interpretation, devised for assigning probabilities to the single case. This problem arises with special emphasis in quantum mechanics 'because the $\psi$-function determines the probability of a *single electron* to take up a certain state, under certain conditions' (Popper 1957, p. 66). Propensities are meant to answer the need to 'take as fundamental the *probability of the result of a single experiment, with respect to its conditions*' (*ibid.*, p. 68). Therefore the conditions surrounding experiments are at the core of Popper's propensity interpretation, which does not associate the 'single case' with particular objects, like particles or dice, but rather with the experimental arrangement (or set-up) in which experiments take place. In Popper's words,

> Every experimental arrangement is *liable to produce*, if we repeat the experiment very often, a sequence with frequencies which depend upon this particular experimental arrangement. These virtual frequencies [...] *characterize the disposition, or the propensity*, of the experimental arrangement to give rise to certain characteristic frequencies *when the experiment is often repeated* (*ibid.*, p. 67).

According to this position, probability interpreted as propensity is a property of the experimental arrangement, or of the generating conditions of an experiment, apt to be reproduced over and over again, to form a sequence. Popper's insistence on the fact that propensities are to be detected in a sequence of experiments suggests that in the Fifties he had in mind some sort of 'long run propensity interpretation', or a variant of the frequency theory, obtained by referring probabilities to generating conditions rather than collectives. Passages like the following suggest this kind of interpretation:

> The frequency interpretation always takes probability as relative to a
> sequence which is assumed as given; and it works on the assumption
> that a probability is a *property of some given sequence*. But with our
> modification, the sequence in its turn is defined by its set of *generating
> conditions*; and in such a way that probability may now be said to be a
> *property of the generating conditions* (Popper 1959, p. 34).

According to Popper, this move opens the door to the attribution of probability
to the single case. All that has to be done is to interpret probability as the
disposition of experimental conditions to produce a certain frequency in every
single experiment that is performed:

> This modification of the frequency interpretation leads almost
> inevitably to the conjecture that probabilities are dispositional
> properties of these conditions – that is to say, propensities. This allows
> us to interpret the probability of a *singular* event as a property of the
> singular event itself (*ibid.*, p. 37).

As illustrated by these passages, Popper oscillates between two different
standpoints, namely a 'long run' and a 'single case' propensity interpretation.
This was pointed out by Gillies, who claims that 'Popper's original propensity
theory was, in a sense, *both* long run *and* single case' (Gillies 2000, p. 126).

The propensity theory was resumed by Popper in *Postscript to the Logic of
Scientific Discovery* (1982), especially volumes I: *Realism and the Aim of
Science*, and III: *Quantum Physics and the Schism in Physics*. There the
propensity interpretation is presented as a variant of the classical theory of
probability, not so much in Laplace's version, but according to the 'measure-
theoretical approach of Cantelli, Kolmogorov, Wald, Church and J.L. Doob'
(Popper 1983, p. 374). Though this approach, which is called by Popper 'neo-
classical', does not imply any interpretation of probability, it is taken to suggest
'an interpretation that *attributes probabilities to single occurrences of events*,
to be *tested* by frequencies within sequences of repetitions of the event in
question; that is to say, it suggests the propensity interpretation of probabilities'
(*ibidem*).

Propensities are defined as 'weighted possibilities' and seen as measurable
expressions of the tendency of a possibility to realize itself upon repetition
(Popper 1982, p. 70). As before, propensity is described as a 'new physical (or
perhaps metaphysical) hypothesis' analogous to Newtonian forces (Popper

1983, p. 360). But now the emphasis is on single arrangements, rather than on sequences of generating conditions, for Popper claims that 'every experimental arrangement (and therefore *every state of a system*) generates propensities' (*ibidem*). In some cases, namely when they are referred to mass phenomena or to repeated experiments, propensities can be measured by means of frequencies. In other cases, they cannot be measured at all, they can only be estimated 'speculatively' (Popper 1990, p. 17). In all cases, statements about propensities, like all other scientific hypotheses, have to be testable. In Popper's words: 'To make a statement about probability is to propose a *hypothesis*. [...] In proposing this hypothesis, we can make use of all sorts of things – of past experience, or of inspiration: it does not matter *how we get it*; all that matters is *how we test it*' (Popper 1957, p. 66). In other words, the acceptability of such statements depends on their testability.

This raises the question of how probabilistic statements are tested. Popper distinguishes between 'probability statements', seen as statements about frequencies in *virtual* sequences of experiments, and 'statistical statements', or statements of relative frequency, seen as statements about frequencies in *actual* sequences of experiments (1982b, p. 70). As already seen, according to the propensity theory probability statements are assignments of 'weighted possibilities', and the weights attached to such possibilities are measures of '(conjectural) virtual frequencies, *to be tested by actual statistical frequencies*' (*ibidem*). Popper illustrates this by means of the following example:

> If we have a large die containing a piece of lead whose position is adjustable, we may *conjecture* (for reasons of symmetry) that the weights (that is, the propensities) of the six possibilities are *equal* as long as the centre of gravity is kept equidistant from the six sides, and that they become *unequal* if we shift the centre of gravity from this position. For example, we may increase the *weight of the possibility* of 6 turning up by moving the centre of gravity away from the side showing the figure '6'. And we may here interpret the word 'weight' to mean 'a measure of the propensity or tendency to turn up upon repetition of the experiment'. More precisely, we may agree to take as our measure of that propensity the (virtual) *relative frequency* with which the side turns up in a (virtual, and virtually infinite) sequence of repetitions of the experiment (Popper, 1982b, p. 70).

Here the 'weight' represents the measure of the propensity of obtaining a given result upon repetition of an experiment, a measure which is assumed to coincide with the virtual frequency of that result in a virtually infinite sequence of repetitions of the same experiment. Such a conjecture will be expressed by a probability statement that can be tested by performing a real sequence of experiments. The latter will lead to a statistical statement expressing the frequency thus obtained, to be then compared with the one that has been conjectured. Probability statements expressing propensities are therefore tested by means of observed frequencies. Their acceptability depends on the possibility of performing a test of this kind.

In the Eighties, the propensity interpretation was regarded by Popper as providing a straightforward solution to the main problems of both the frequentist and classical interpretations of probability. On the one hand, propensity theory was said to solve the problem of the single case posed by frequentism, on the other it was granted the merit of avoiding determinism, to which the classical interpretation is committed. Popper does not mention the subjectivist interpretation of probability worked out by Ramsey and de Finetti; instead he associates subjectivism with the classical interpretation, and says that 'the subjective interpretation was adopted (especially by Laplace) *because* it was logically compatible with determinism, and because it was apparently the *only* one that was compatible' (*ibid.*, p. 110). As we will see, this claim would not hold for the subjectivism of authors like Ramsey and de Finetti.

## A world of propensities

The notion of propensity gained increasing importance in Popper's philosophy. So much so that in the last years of his life Popper put it at the core of a metaphysical programme aimed at combining in a unitary framework with a strong realistic and indeterministic slant all sorts of causal tendencies operating in the world, from micro-physics to human action[6]. In his essay *A World of Propensities* (1990), the propensity interpretation is endowed with a 'cosmological significance', based on the claim that 'we live in *a world of propensities*, and [...] this fact makes our world both more interesting and more homely than the world as seen by earlier states of the sciences' (Popper 1990, p. 9). Once more, Popper reaffirms the objective and realistic character of propensities, which are physical entities akin to physical forces. The propensity interpretation

---

[6] The development of Popper's propensity theory from the Fifties to the Nineties, in connection with its realistic and indeterministic implications, is surveyed in Runde (1996).

*is an objective interpretation of the theory of probability*. Propensities, it is assumed, are not mere possibilities but are physical realities. They are as real as forces, or fields of forces. And vice versa: forces are propensities. They are propensities for setting bodies in motion. Forces are propensities to accelerate, and fields of forces are propensities distributed over some region of space and perhaps changing continuously over this region (like distances from some given origin). Fields of forces are fields of propensities. They are real, they exist (*ibid.*, p. 12).

After a few pages, we read that 'like Newtonian attractive forces' propensities 'are invisible, and, like them, they can act: they are *actual*, they are *real*' (*ibid.*, p. 18).

According to the perspective taken by Popper in 1990, propensities are much more than objective chances. Not only do they inspire an objective theory of probability, but underpin an indeterministic conception of the world. In Popper's words:

> with the introduction of propensities, the ideology of determinism evaporates. Past situations, whether physical or psychological or mixed, do not determine the future situation. Rather, they determine changing *propensities that influence future situations without determining them in a unique way*. And all our experiences – including our wishes and our efforts – may contribute to the propensities, sometimes more and sometimes less, as the case may be (*ibid.*, p. 17-8).

Propensities are now conceived as properties of '*the whole physical situation*', having an indeterministic and situational character, meant to embrace in the same picture probabilistic tendencies of all kinds, ranging from physics and biology to the motives of human action. Popper's world of propensities is a world in constant evolution:

> in our real world, the situation and, with it, the possibilities, and thus the propensities, change all the time. They certainly may change if we, or any other organism, *prefer* one possibility to another; or if we *discover* a possibility where we have not seen one before. Our very understanding of the world changes the conditions of the changing world; and so do our wishes, our preferences, our motivations, our hopes, our dreams, our phantasies, our hypotheses, our theories. [...] All this amounts to the fact that *determinism is simply mistaken*: all its

traditional arguments have withered away and indeterminism and free will have become part of the physical and biological sciences. [...]

This view of propensities allows us to see in a new light the processes that constitute our world: the world process. The world is no longer a causal machine – it can now be seen as a world of propensities, as an unfolding process of realizing possibilities and of unfolding new possibilities (*ibidem*).

## 5.3 After Popper

### Single-case and long-run propensity theories

After Popper, the propensity theory of probability gained increasing popularity among philosophers of science, and provoked a debate that is much too vast to be examined in full detail. Broadly speaking, the supporters of the propensity theory fall into two groups, according to whether they endorse a single-case or a long-run version of it. Single-case propensity theories have been put forward by such authors as Ronald Giere, Hugh Mellor, David Miller and James Fetzer, whereas others, including Donald Gillies, have moved closer to the long-run propensity interpretation[7].

Giere's approach 'makes no essential reference to any sequence (virtual or real)' (Giere 1974, p. 473) and takes propensity as a theoretical property of chance set-ups, to be interpreted as a disposition to produce specific outcomes on particular trials with a certain strength. Giere regards such properties as non-Humean, and denies the existence of a structural link between propensities and frequencies. As he puts it: 'there is no direct logical connection between single-case propensities and relative frequencies, not even "in the limit"' (*ibid.*, p. 478). Still, he has to admit that 'relative frequencies may provide *evidence* for propensity hypotheses' and even that 'in the absence of a well-developed theoretical background, observed relative frequencies may provide the only evidence for propensity statements' (*ibidem*). Giere's view is openly metaphysical, for it involves a strong metaphysical presupposition, which

---

[7] The number of authors who have embraced some version of the propensity theory is big enough to discourage a more detailed account of the debate on this matter. Among those whose position is not discussed here, we mention Levi (1967) and Hacking (1965) and (1980).

amounts to the claim not only that 'there are physical possibilities in nature, but further that nature itself contains innate tendencies towards these possibilities, tendencies which have the logical structure of probabilities' (Giere 1976, p. 348). Giere's theory embodies an argument to the effect that propensity structures which are present in nature generate probability spaces[8].

A slightly different version of single-case propensity theory is put forward by Hugh Mellor, whose approach strays from Giere's in at least one important respect. This amounts to the fact that, while propensities are seen by Mellor as dispositions to display chance distributions over the possible results of a certain trial, they should not be confused with such distributions, because they do not admit degrees in themselves. In other words, propensity is a disposition to display a chance distribution in every trial, rather than a graded disposition, as Giere would have it. In Mellor's words:

> a disposition may in general be characterized as the feature of a trial that constitutes its display. That feature is often an event, some characteristic result of the trial. But a statement ascribing the disposition does not entail that the characteristic event ever occurs because it does not entail that the disposition is ever displayed. [...] a propensity statement, being a disposition statement, further does not entail that the trial occurs (Mellor 1971, p. 71).

It is noteworthy that for Mellor propensities are not probabilities, and therefore a demonstration that propensity can be linked to a probability function is not needed. In this spirit, he claims to put forward a propensity *theory*, rather than a propensity interpretation of probability[9].

David Miller, who worked closely with Popper and with whom he co-authored some papers against inductive method[10], embraces a single-case propensity view, emphasizing the importance of taking into account a larger framework than that consisting in the experimental conditions surrounding an event. Miller believes that propensities should be made to depend on no less than the complete state of the universe at the time the considered event takes place. In his words,

---

[8] See Giere (1976).

[9] For a more detailed, critical account of Mellor's propensity theory see Salmon (1979b).

[10] See, among others, Popper and Miller (1987), including extensive references to the relevant literature.

According to the propensity interpretation, in the strong form in which I recommend it here, what we call *the* probability of an outcome is an absolute (though not as inexorably permanent) a characteristic as is the age of a person: it is relative only to the unique situation of the world (or the causally operative part of the world) at the time (Miller 1994, p. 183).

Taking for instance the case of life expectancy, one can say that the probability of my survival a certain number of years from today depends on myriad factors, some of which are directly related to my own particular situation – such as my activities, lifestyle and genetical make-up – while others are external to this situation. Hence, for Miller the assertion 'that the probability of my death within a year is a property of my local surroundings [...] is mistaken; the present work, or enthusiasm for work, of distant medical researchers is an important determinant too' (*ibid.*, p. 185). He concludes that

Propensities are not located in physical things, and not in local situations either. Strictly, every propensity (absolute or conditional) must be referred to the complete situation of the universe (or the light-cone) at the time (*ibidem*).

Miller's propensionism is close to the position upheld by Popper in the Eighties and Nineties, unless propensities are not said to generate frequencies, being rather a direct expression of objective single-case probabilities. As maintained by Miller: 'In the propensity interpretation, the probability of an outcome is not a measure of any frequency, but [...] a measure of the inclination of the current state of affairs to realize that outcome' (*ibid.*, p. 182). A trait shared by Miller and the later Popper is that propensities are metaphysical entities, which are not always measurable. This may indeed sound puzzling, as seems to be the case of Donald Gillies, who observes that 'The main problem with the 1990s views on propensity of Popper and Miller is that they appear to change the propensity theory from a scientific to a metaphysical theory' (Gillies 2000, p. 127).

A further theorist of single-case propensity is James Fetzer, who severs the link between propensity and frequency characterizing the long-run propensity interpretation, to establish instead a causal connection between them. According to Fetzer, propensity is a dispositional property displayed by a chance set-up in the single case, characterized by a strength that does not change from trial to trial. The truth of a probabilistic statement, then, 'entails the existence of some dispositional tendencies (if there are any chance set-ups of that kind) but no infinite sequences – indeed, it entails not even one actual trial' (Fetzer 1974a, p. 391). Within Fetzer's perspective, emphasis is switched from the state of the universe at a given time to the set of nomically (or causally) relevant conditions framing events. Fetzer's move is regarded by Gillies as problematic, pretty much in the same sense as Miller's:

> If, as Fetzer suggests, we ascribe propensities to a complete set of (nomically and/or causally) relevant conditions, then in order to test a conjectured propensity value we must make a conjecture about the complete list of the conditions which are relevant. This necessary conjecture might often be difficult to formulate and hard to test, thereby rendering the corresponding propensities metaphysical rather than scientific. Once again then I have a doubt whether single-case propensities give an appropriate analysis of the objective probabilities which appear in the natural sciences (Gillies 2000, p. 128).

This passage suggests that single-case propensities run into a reference class problem, similar to that affecting also Reichenbach's attempted solution to the problem of the single case. As observed in Chapter 4, it is always problematic – and in some cases even impossible – to identify the proper reference class for particular events. Even granted that that were possible, one might object to the uncritical use of a reference class based on available statistics on the basis of various considerations, like the fact that there could be further evidence of a qualitative kind, or that the event could be classified according to different criteria[11].

To avoid the problematic aspects of single-case propensity theories, Gillies opts for a long-run propensity view. Such an approach is closer to that put forward by Popper in the Fifties, and is set forth as a variant of frequentism, which shares a number of features of von Mises' approach – taken as the standard frequency interpretation. As described by Gillies, the main traits in common with his own propensity view and von Mises' frequentism amount to the following:

> (1) probability theory is a mathematical science like mechanics or electromagnetic theory; (2) this science deals with random phenomena to be found in the material world; (3) its axioms are confirmed empirically; and (4) probabilities exist objectively in the material world, like masses or charges, and have definite, though perhaps unknown,

---

[11] See the discussion of the problem of the reference class in connection with single-case propensities in Gillies (2000), pp. 119-25.

values (*ibid.*, p. 137).

Gillies' propensity approach strays from von Mises' frequency interpretation in associating probabilities with repeatable conditions instead of collectives, and in taking a non-operationalist perspective. This is a point stressed by Gillies, who introduces propensities as 'theoretical concepts', which 'are not defined in terms of observables' (*ibidem*). According to Gillies, this propensity interpretation should be retained for application to the natural sciences, and combined with the subjective notion of probability – more suitable for application to the social sciences – within a broader perspective, which qualifies as genuinely pluralistic[12].

## Humphreys' paradox

A problem for the propensity theory pointed out by Paul Humphreys has raised important issues and provoked a wide debate. In 'Why Propensities cannot be Probabilities' (1985), Humphreys argues that single-case propensities do not cover all of the uses of probability and, more particularly, that they are unsuitable to represent inverse probabilities. A direct consequence of this is that Bayes' theorem is inapplicable to propensities. The problem originates from the dispositional character of propensities, defined as tendencies to produce certain outcomes. Such a dispositional character confers to propensities a peculiar asymmetry, which clashes with the symmetry characterizing probabilities, as described by the probability calculus. As Humphreys puts it,

> the properties of conditional propensities are not correctly represented by the standard theory of conditional probability; in particular any result involving inverse probabilities, including Bayes' theorem, will, except in special cases, give incorrect results (Humphreys 1985, p. 559).

Noticeably, the asymmetry of propensities is not just temporal, being rather causal in character. This has persuaded various authors to regard the notion of propensity as more appropriate to represent causal tendencies than probabilities. This is the case of Wesley Salmon, who in his 'Propensities: a Discussion Review' (1979) reaches a similar conclusion, after taking seriously the problem raised by Humphreys. Referring to the latter, he observes that

---

[12] See Gillies (2000) for more details.

Given suitable 'direct' probabilities we can, for example, use Bayes' theorem to compute the probability of a particular cause of death. Suppose we are given a set of probabilities from which we can deduce that the probability that a certain person died as a result of being shot through the head is 3/4. It would be strange, under these circumstances, to say that this corpse has a propensity (tendency?) of 3/4 to have had its skull perforated by a bullet. Propensity can, I think, be a useful causal concept in the context of a probabilistic theory of causation, but if it is used in that way, it seems to inherit the temporal asymmetry of the causal relation (Salmon 1979b, pp. 213-4).

Salmon does not deny the usefulness of the notion of propensity. On the contrary, he believes that the asymmetry of this notion, its dispositional character, and its direct bearing on the single-case recommend its adoption for the sake of causal analysis, whereas the same features make it unsuitable to the interpretation of probability. In Salmon's words:

Propensities are causal probabilities, and, as such, they play an indispensable role in the probabilistic causal mechanisms of the universe. However, not all probabilities are causal in that way, and so it is a mistake, in my opinion, to try to *define* 'probability' in terms of propensities (Salmon 1984, p. 205).

As to the interpretation of probability, Salmon sticks to frequentism, and more precisely to Reichenbach's version of it, and adopts the notion of propensity as an ingredient of a notion of causality taken in a probabilistic sense[13].

While the vast discussion on Humphreys' problem cannot be given here[14], it is worth noticing that Gillies regards it as a definite advantage of his own version of propensity theory that it does not run into difficulties of that kind. To argue this, Gillies considers an example, which is actually very close to the one examined by way of an illustration of Bayes' theorem in Chapter 2. The example concerns two machines producing frisbees, the first of which produces 800 pieces per day, with 1% defective, the second produces 200 pieces per day with 2% defective. At the end of each day a frisbee is selected at random from 1,000 produced by the two machines. Posing $D$ = the selected frisbee is defective; $M$ = it was produced by machine 1; and $N$ = it was produced by

---

[13] For more details on Salmon's sophisticated theory of probabilistic causality, see Salmon (1984) and (1998).

[14] More on the issue can be found in Milne (1986) and McCurdy (1986).

machine 2, one can consider the conditional probabilities $P(D \mid M)$ and $P(M \mid D)$. We know that $P(D \mid M) = 0.01$. A simple application of Bayes' theorem allows us to calculate $P(M \mid D)$:

$$P(M \mid D) = [P(D \mid M) P(M)] / [P(D \mid M) P(M) + P(D \mid N) P(N)]$$
$$= (0.01 \times 0.8) / (0.01 \times 0.8 + 0.02 \times 0.2) = 8/12 = 2/3.$$

The question is: does it make sense to interpret these probabilities in terms of propensities? In terms of single-case propensities, $P(D \mid M)$ is not problematic, because it can simply be taken as the propensity of machine 1 to produce a defective frisbee. Things are different as with the interpretation of $P(M \mid D)$, which is far from straightforward, for it would mean the propensity of the defective frisbee picked at random at the end of the day to have been produced by machine 1. No doubt, this sounds awkward. In Gillies' words:

> If we think of propensities as partial causes, this becomes the following. The drawing of a defective frisbee in the evening is a partial cause of weight 2/3 of its having been produced by machine 1 earlier in the day. Such a concept seems to be nonsense, because by the time the frisbee was selected, it would either definitely have been produced by machine 1 or definitely not have been produced by that machine. [...] Obviously examples of this sort pose a problem for the propensity view of probability (Gillies 2000, p. 131).

On the contrary, Gillies believes that the adoption of the long-run version of propensity theory helps to circumvent the problem, because in that case the value of $P(M \mid D)$ is taken to represent the propensity to produce defective pieces with a certain frequency. Thus:

> If propensities are propensities to produce long-run frequencies, then a propensity of 2/3 makes perfect sense, even though we know that in each individual case the result has been definitely determined as either $M$ or $N$ by the time the selected frisbee is examined and found to be defective (*ibid.*, pp. 132-3).

On such a propensity account, propensities do not receive a direct causal meaning, being rather the expression of statistical correlations. Given a sudden fall of the barometer, there is a high propensity of rain, but this does not involve a causal connection between the two.

In broad terms, one can say that there is a tension between the probabilistic and the causal interpretation of propensity. On the one hand, taking propensities as causal tendencies to exhibit a certain result in the single-case clashes with their interpretation as (standard) probabilities. This consequence has been accepted, among others, by Fetzer, who takes propensities to be non-standard probabilities. If, on the other hand, propensities are taken as probabilities, as done by Gillies, the natural move seems to interpret them as long-run propensities, while divorcing them from a direct causal interpretation. Salmon's position is still different, for he does not take propensities to be probabilities, but causal properties, which only apply when knowledge of causal mechanisms and processes allows causal talk with reference to single events[15].

## Propensity as an ingredient of the description of chance phenomena

A still different approach to the notion of propensity – which is gaining increasing popularity within recent literature in philosophy of science[16] – regards it as a useful ingredient of the description of chance phenomena, not to be necessarily identified with probability. A perspective of this kind, which strays in important respects from all of the other accounts of propensity, has been put forward by Patrick Suppes. In a number of papers merged in his recent book *Representation and Invariance of Scientific Structures* (2002), Suppes favours the adoption of the notion of propensity, taken as a physical dispositional property useful for the purpose of analysing the structure of a number of phenomena. Suppes does not take sides either with the long-run or single-case propensity theories, nor does he aim at a philosophical characterization of the notion of propensity. Furthermore, he does not regard propensities as expressing probabilities, in which connection he claims that he 'never for a moment thought that propensities did have the properties of probability' (Suppes 2002, p. 223). Quite the contrary, for Suppes propensity is an extra-ingredient, with respect to the probabilities involved in the description of certain phenomena. Propensities are rather taken to 'provide the ingredients out of which probabilities are constructed' (*ibid.*, p. 209-10). More precisely, propensities confer an objective meaning to the probabilities involved in the description of a whole array of phenomena.

In this spirit, Suppes works out representation theorems for various kinds of chance phenomena, including coin tossing, radioactive decay, and a

---

[15] See Footnote 13.

[16] Evidence to this claim is offered – among other things – by the recent publication of a double issue of *Synthèse*, CXXXII (2002), nos. 1-2, devoted to 'Propensities and Probabilities'.

phenomenon belonging to psychology, namely response strength. In all such cases, plus a version of the three-body problem that exhibits what he calls a 'propensity for randomness', in order to obtain a representation of the phenomena at hand, one has to give structural axioms, built on information which is not purely probabilistic in character. Representation of decay, for example, is obtained by means of a 'waiting-time axiom' which 'is a structural axiom that would never be encountered in the standard theory of subjective probability as a fundamental axiom. It is an axiom special to certain physical phenomena. It represents, therefore, a qualitative expression of a propensity' (*ibidem*). In the course of his analysis of radioactive decay, Suppes calls attention to the fact that 'the probabilities we obtain from the representation theorem are not unique but are only unique up to fixing the decay parameter. Again, this is not a subjective concept but very much an objective one' (*ibidem*). One might say that Suppes' notion of propensity provides a tool for representing chance phenomena, whose description involves 'objective' considerations that put some constraint on the estimation of probability.

Suppes, who upholds a typically pluralistic attitude towards philosophy of science in general[17], does not claim that the notion of propensity covers all uses of objective probability. In fact he maintains that 'it is a virtue of propensity theory to provide a particular causal view of probability, which will qualify as a special kind of objective probability' (*ibid.*, p. 221). To instantiate this claim, he adds that 'there is no reason in principle that accounts of probability or of randomness in terms of complexity can always be reduced to accounts in terms of propensity' (*ibidem*). Thus for Suppes it is not even worth trying to give a general account of propensity: 'there is not going to be a general account of propensities, just as there will not be a deep general account of the physical properties of the world' (*ibid.*, pp. 224-5). Besides, he is convinced that the notion of propensity – like that of cause – does not appear in the formulation of worked out scientific theories. The fruitfulness of such notions shows 'in informal talk surrounding the theory and often in informal descriptions of experiments or of observational procedures' (*ibid.*, p. 221).

An appeal to propensities to represent objective probabilities is also made by the philosopher of physics Abner Shimony, whose approach seems to share Suppes' pluralism, at least in this connection. Shimony believes that, in order to properly describe various kinds of phenomena, a mixture of epistemic – or 'personal', as he calls it – and objective probability is called for. He further

---

[17] More on Suppes' philosophy will be said in Chapter 7, Section 4.

believes that the objective probabilistic component is best represented in terms of propensity. Here propensity is not taken as a univocal notion. On the contrary, Shimony speaks of a 'family' of propensities, representing probabilities that are objective characteristics of various systems. The 'primary' representatives of such a family are, on the one hand, 'classical propensities' encountered in classical physical systems, and, on the other, 'quantum propensities' belonging to quantum systems. As Shimony puts it,

> Quantum mechanical probabilities have three features in common with the classical propensities: (a) objectivity in contrast with personal or subjective probability, (b) probabilities are derived from the constitution of the system, and (c) probability can be attributed to the individual situation and then derivatively to the ensemble, in contrast to the frequency theory of probability; but they disagree with classical propensities regarding two features: (d) the background of deterministic physical laws, and (e) a moderate appeal to epistemic probability (Shimony, *unpublished manuscript*).

As to points (a), (b) and (c), there is not much to add to Shimony's account, because such points just reflect the reasons why, from Popper to Gillies, the propensity interpretation is taken to represent physical probabilities better than the frequency interpretation. Point (d) marks a disagreement between Shimony and Popper, because the latter denies that there is a substantial difference between classical and quantum physics as regards determinism, and includes both of them in a broadly indeterministic perspective, while Shimony argues for a neat differentiation of the two cases. To his eyes, quantum mechanical systems are in fact characterized by a peculiar objective 'indefiniteness', reflected by Heisenberg's notion of 'potentiality', meant

> to characterize a property which is objectively indefinite, whose value when actualized is a matter of objective chance, and which is assigned a definite probability by an algorithm presupposing a definite mathematical structure of states and properties (Shimony 1999, p. 6).

Propensity proves fruitful precisely in connection with the representation of Heisenberg's notion of potentiality, which reflects the idea that the Copenhagen interpretation of quantum theory revolves around 'the transition from the possible to the actual' (Heisenberg 1959, 1990, p. 131). Moreover, 'the atoms or the elementary particles [...] form a world of potentialities or possibilities rather than one of things of fact' (*ibid.*, p. 174). In fact, according to quantum theory it is only during the act of observation that the passage from the possible to the real can take place. Point (e) in Shimony's list concerns the role of personal probability in the description of physical systems. Personal probability judgments are needed in the case of both classical and quantum systems, though in a slightly different sense. In the first case, a 'weak judgment' regarding initial conditions is required, while in the second epistemic probability comes into play with the acceptance of quantum theory itself, or a particular version of it[18].

Having compared classical and quantum systems and argued for the fruitfulness of the adoption of the notion of propensity in both cases, Shimony contends – without developing the matter much further – that propensity is liable to prove useful in other fields, such as biology, and more specifically population genetics.

Neither Suppes nor Shimony seem to be interested in working out a philosophical theory of propensity, of the kind put forward by Miller, Giere, or Fetzer, or to take sides on the issue of single-case versus long-run interpretations. Their appeal to propensity is limited to an analysis of cases involving some notion of chance, or objective probability, taken with reference to some sort of system, and bearing directly on the single case. Their work is more evidence of the popularity the notion of propensity has gained among philosophers of science in recent years.

Not all the views aired in the debate on propensities, however, have been favourable. Some doubts on the suitability of this notion to represent physical, and especially quantum probabilities have also been raised by a number of authors, including Peter Clark, Stephen Leeds and Jenann Ismael. On the basis of various arguments, which are much too complex to be analysed in detail, these authors come to a sceptical conclusion on Popper's pretence that the propensity theory 'takes the mystery out of quantum theory'. As Clark puts it, 'Neither in application to statistical mechanics nor the quantum theory does the propensity theory of probability resolve the outstanding conceptual problems of those theories' (Clark 1995, p. 161). Taking it for granted that frequentism leaves crucial problems open, and sharing Clark's scepticism towards the propensity theory, Leeds and Ismael opt for epistemic probability. Thus Ismael claims that 'chance is just epistemic probability of a particular sort' (Ismael 1996, p. 89), and Leeds takes a similar position, by claiming: 'the contribution

---

[18] Shimony himself gives an overview of the various interpretations of quantum mechanics existing in the literature in Shimony (1999).

of chances is, I suggest, to fix our subjective probabilities' (Leeds 1984, p. 573). In Chapter 7 we will return to the issue of accounting for chance and physical probability within the framework of subjective probability.

## 5.4 Digression on chance and randomness

At this point of our presentation, it does not seem out of place to spend a few words on the notion of *chance*, which appears to be strictly related to the empirical meaning of probability. Despite such a relationship, which now seems obvious, chance and probability have a different history – as testified by the fact that the notion of chance has been part of western thought since its early days, whereas the notion of probability in its modern meaning took shape much later. According to Hacking, the emergence of probability was fostered by the encounter between probability and 'numerical ideas of randomness' that occurred around the mid-seventeenth century (Hacking 1975, p. 18).

### Historical remarks

The writings of many authors of antiquity contemplate a notion of chance referring essentially to what is fortuitous, contingent, not amenable to the causal net underlying phenomena, and therefore unpredictable. Evidence to this effect is to be found in the work of Democritus, Aristotle, Epicurus, Cicero and Boethius, to mention just a few. According to a widespread tendency, chance is produced by the concurrence of independent causal chains. For instance, Antoine Augustin Cournot (1801-1877), a great supporter of the fruitfulness of applying mathematical techniques to economics and social affairs[19], writes in his *Exposition de la théorie des chances et des probabilités* (1843) that 'facts resulting from the combination or intersection of phenomena belonging to independent chains in the causal network are called *fortuitous*, or deriving from *chance*' (Cournot 1843, 1984, p. 55). A similar stance is taken by John Stuart Mill, who in his *System of Logic* claims that

> we may say that two or more phenomena are conjoined by chance, that they coexist or succeed one another only by chance; meaning that they are in no way related through causation; that they are neither cause and effect, nor effects of the same cause, nor effects of causes between

---

[19] On Cournot see the *Revue de Métaphysique et de Morale* XIII (1905), no. 3, entirely devoted to his work.

which there subsists any law of coexistence, nor even effects of the same collocation of primeval causes (Mill 1843, 1973, 1996, vol. 7, pp. 526-7; Book III, Chapter xvii, § 2).

The conception of chance embraced by these authors was indeed very influential, and imbued a whole tradition of thought. As we saw, it was somehow retained also by John Venn, who tried to combine it with his frequentist view of probability, which implied a shift of attention from individuals to sequences, in view of the determination of probabilities. A more radical attitude was taken by von Mises, who – in line with the frequency standpoint – concentrated on random sequences. We will soon return to von Mises' notion of randomness.

Other authors emphasise the epistemic character of chance, which they impute to our ignorance of causes. Among them Paul Henri Thiry d'Holbach (1723-1789), one of the major exponents of the French Enlightenment, contributor of a number of articles for the *Encyclopédie* of Denis Diderot and Jean d'Alembert, and a convinced supporter of atheistic materialism. In his *Système de la nature, ou des lois du monde physique et du monde moral* (1770) d'Holbach says that 'man uses the word *chance* to cover his ignorance of those natural causes which produce visible effects, by means of which he cannot form an idea; or that act by a mode of which he does not perceive the order' (d'Holbach 1770, English edition 1970, p. 37).

A similar perspective is embraced by Laplace, who – as we saw in Chapter 3 – regards chance as resulting from our ignorance of causal links holding between phenomena. Such a conception of chance is rooted in Laplace's determinism, and goes hand in hand with his epistemic view of probability.

## Poincaré's views on chance

A decisive turn in the literature on chance was made by the French mathematician Henri Poincaré (1854-1912), who gave substantial contributions to various branches of mathematics and geometry, as well as celestial mechanics and mathematical physics. Extremely prolific both as a scientist and as a writer, in addition to about five-hundred scientific papers, Poincaré left a number of philosophical essays, including the books *La science et l'hypothèse* (1902), *La valeur de la science* (1905), and *Science et méthode* (1908), and the writings collected in *Dernières pensées* (1912)[20]. He also wrote a treatise on probability, entitled *Calcul des probabilités* (1912). Poincaré wants to go beyond the intuitive notion of chance as absence or ignorance of causes, and attempts to analyse the nature of chance phenomena themselves. He starts by distinguishing, among those phenomena whose causes are unknown, 'fortuitous phenomena', to which the calculus of probabilities applies, and those phenomena which do not qualify as fortuitous, 'about which we can say nothing, so long as we have not determined the laws that govern them' (Poincaré 1908, English edition 1914, 1996, p. 66). Then Poincaré proceeds to characterize the distinctive features of fortuitous phenomena.

A typical case in point obtains when very small causes, or very small differences in the initial conditions of a phenomenon (so small that they cannot be registered) produce macroscopic differences in the final result. In such cases, prediction of the final happening becomes impossible, and a fortuitous phenomenon obtains. Phenomena of this kind are easily found in meteorology, and in general in all situations characterized by an unstable equilibrium of their elements. A familiar example is that of roulette, described by Poincaré as follows:

> Imagine a needle that can be turned about a pivot on a dial divided into a hundred alternate red and black sections. Clearly, all depends on the initial impulse we give to the needle. I assume that the needle will make ten or twenty revolutions, but it will stop earlier or later according to the strength of the spin I have given it. Only a variation of a thousandth or a two-thousandth in the impulse is sufficient to determine whether my needle will stop at a black section or at the following section, which is red. These are differences that the muscular sense cannot appreciate, which would escape even more delicate instruments. It is, accordingly, impossible for me to predict what the needle I have just spun will do, and that is why my heart beats and I hope for everything from chance. The difference in the cause is imperceptible, and the difference in the effect is for me of the highest importance, since it affects my whole stake (*ibid.*, pp. 69-70).

---

[20] Poincaré's scientific works are collected in Poincaré's *Oeuvres* (1928-56), which include a biography of him by Gaston Darboux (in volume 2) and a number of centenary lectures on his life and work (in volume 10). A collection of essays intended for philosophers is to be found in the *Revue de Métaphysique et de Morale*, vol. XXI (1913), Supplement to no. 5.

*Instability* is therefore a distinctive feature ascribed by Poincaré to chance phenomena.

Other phenomena qualify as fortuitous not so much for their instability, as their *complexity*. This holds in the first place for the phenomena described by the kinetic theory of gases. Here there is a further element determining the unpredictability of molecular behaviour, namely the immense number of molecules contained in a receptacle full of gas, which results in an enormous number of collisions, and consequently in an overly complex ensemble of causes. The same happens, says Poincaré, when two liquids or two fine powders are mixed, or when a pack of cards is shuffled long enough, for

> if the player shuffles the cards, there will be a great number of successive permutations, and the final order which results will no longer be governed by anything but chance; I mean that all the possible orders will be equally probable. This result is due to the great number of successive permutations, that is to say, to the complexity of the phenomenon (*ibid.*, p. 74).

Instability and complexity therefore turn out to be the chief characteristics of fortuitous phenomena, which determine their unpredictability. In order to qualify as produced by chance, phenomena have to be such that the final situation is independent of the initial one. If instead some feature of the initial conditions remains unchanged, the final state is no longer independent of the initial one, and the phenomenon cannot be called fortuitous.

## The riddle of randomness

Poincaré's analysis of chance inspired Richard von Mises, who agreed with the French mathematician that Laplace's concept of chance should be rejected, and shared his conviction that chance should be given a conceptual priority with respect to probability, in the sense that probability applies to chance phenomena. As we saw, von Mises, who concentrated on collectives rather than on particular phenomena, restated the problem as that of defining a notion of randomness to be applied unrestrictedly to probabilistic sequences. As observed in the last chapter, von Mises' notion of randomness was shown to be beset with difficulties, the most serious of which amounts to the impossibility of producing a demonstration that there are sequences which are random in von Mises' sense, except for the trivial case of sequences whose attributes have probability 0 or 1. Among the authors who have tried to cope with this and other difficulties, one should mention the statistician Abraham Wald and the logician Alonzo Church. In an attempt to improve on von Mises' definition, Wald suggested that the notion of effective calculability be adopted to define random sequences and proved that, if an effectively calculable set of place selections is taken, the set of sequences insensitive to place selections associated with a given probability value, other than 0 or 1, is a continuum[21]. Church strengthened this condition by requiring that the notion of effective calculability be replaced by that of recursive enumerability[22]. Though the work of Wald and Church was widely recognized to be a decisive improvement over von Mises' definition, the notion of randomness that emerged was still considered far removed from scientific practice[23].

This led various authors to explore alternative approaches. Among them, Per Martin-Löf and Claus Peter Schnorr worked out a theory of randomness based on the notion of statistical test, while Andrej Kolmogorov built on the idea that randomness can be accounted for in terms of complexity[24]. Following this path, the literature on randomness eventually returned to the notion of complexity, that was already associated with chance by Poincaré. Within such a literature the conviction has steadily grown that the absolute, unrestricted notion of randomness von Mises conceived of should rather leave room for a weaker, relative notion. Indeed, this path was taken in the first place by Reichenbach, whose position has already been considered.

Popper proceeded along similar lines, by introducing 'chance-like' sequences defined over finite segments, as to which he demonstrated that they can be indefinitely extended through a transition to infinity[25]. However, in the Eighties Popper restated the problem in different terms, in tune with his adoption of the propensity interpretation of probability. According to the perspective taken in those years, the propensity theory is seen as a variant of the 'neo-classical' measure-theoretical approach, and a solution to the problem of randomness open by von Mises is proposed, based on a result due to Joseph L. Doob. In Popper's words, this result says that

under the assumption that the sequences in question consist of

---

[21] See Wald (1937).

[22] See Church (1940).

[23] For an excellent survey of the problems raised by von Mises' theory of randomness, see Martin-Löf (1969).

[24] For a review of the theories advanced by these authors see van Lambalgen (1987). For more details on Kolmogorov's approach see von Plato (1994).

[25] See Popper (1934, English edition 1959, 1968), Chapter 8, and 'Appendix *vi': 'On objective disorder or randomness'.

*independent* events, every gambling system fails in *almost all* sequences (in all sequences of independent events, with the exception of a set of measure zero) (Popper 1983, p. 372).

Since the measure-theoretical approach takes measure zero to mean zero probability, but not impossibility, this account of randomness is weaker than von Mises'. However, Popper regards it as sufficiently good for practical purposes. Incidentally, it may be conjectured that von Mises would have disapproved of this way of putting the matter, because he reaffirmed the priority of randomness over probability, whereas independence is a probabilistic notion, so an account of randomness based on independence would have looked circular to him.

A conception more in tune with von Mises' approach based on 'place selections', though aimed at defining a restricted concept of randomness, was worked out by the statistician Maurice George Kendall, who defined randomness relative to a 'selector' – namely 'an infinite series of positive integers ordered according to their magnitude' (Kendall 1941, p. 2) – to be used to pick samples from a given sequence of data. With respect to this approach, von Mises' randomness represents a limiting case, which obtains as the domain of selectors becomes indefinitely larger. Von Mises' unrestricted notion of randomness is considered of no practical importance by Kendall, who, on the contrary, holds that a useful notion of randomness is to be restricted to a finite domain of selectors. The relativity of this notion is emphasized over and over again by Kendall, who claims that 'there is no such thing as absolute randomness, just as there is no such thing as absolute velocity' (*ibid.*, p. 5). He goes so far as to maintain that

> we can never rid ourselves entirely of the possibility that a method of selection may lack randomness, but we can safeguard against this possibility to a great extent. For instance, the Random Sampling Numbers applied to a universe of names in a directory gives us something near certainty [...] that the resulting samples will be random. Furthermore, we can experiment with a method to see if bias has appeared. If it has not, we are justified in expecting that it is random for the class of cases in which it has been tried. Ultimately, however, the assumption of randomness is part of the hypothesis which is being tested (*ibid.*, p. 14).

As mentioned in this passage, in practical research random sampling is obtained by means of various methods, like the so-called 'tables of random numbers' listing numbers that do not exhibit any apparent order, and are therefore unpredictable. For all practical purposes, the use of such tables guarantees that the samples so obtained are random, which allows for the application of a wide range of statistical methods which require the assumption of independence.

The notion of randomness we end up with would be better qualified as 'pseudo-randomness'. In fact, someone who knew how to generate the sequence of random numbers itself, would be in a position to predict it. Kendall observes that randomness, as it occurs within practical research, is itself an assumption made by researchers, who are warned against the danger of transposing random samples from one research framework to another, and are exhorted to bear in mind that 'randomness is relative' (ibidem).

As a matter of fact, also in everyday life we are faced with the relativity of randomness. To exemplify this claim, let me mention a fact that was reported some years ago by an Italian weekly. To promote sales, a biscuit factory launched a 'scratch and win' lottery. A card chequered with squares of different colours was inserted into every packet, to be scratched by consumers. After scratching four black squares at random, consumers finding four equal figures would have won a prize of 5,000 Italian liras. If, scratching one more square, they found a coffer, the prize would have been doubled; but if they discovered a golden coffer they would have been awarded a prize of 3,000,000 liras. Having estimated the probability of finding four equal images as very low, the managers of the factory thought that a small increase in the price of the biscuits would be enough to cover the cost of the lottery. However, one of the inhabitants of Limbadi, a small village near Vibo Valentia, in Southern Italy, noticed that the cards, printed in series, could be distinguished by some tiny clues (like little spots, or different colour shades) left by the printing machine. After comparing a number of cards, the citizens of Limbadi were able to guess to which series each card belonged. By scratching all the squares of one card out of each series, they discovered how the images were distributed among the cards, and this enabled them to devise a winning strategy. Eventually, they got hold of the super-prize, organized a party in the main square, and invited everybody for a big celebration. But while the people of Limbadi had discovered a gambling system, other consumers from all over Italy were just trying their luck scratching the cards at random, and random the lottery was for them. The moral to be drawn from this example seems to be that the same phenomenon can be random in the absence of relevant information, to become predictable in the light of such information. Should we then conclude that chance is a product of human ignorance, as Laplace put it?

## Is chance objective?

To the question: 'Has chance, thus defined so far as it can be, an objective character?' (Poincaré 1908, English edition 1914, 1996, p. 84) Poincaré gave a subtle answer. As we saw, he definitely opposed Laplace's tenet that chance is an epistemic notion, arguing instead that chance arises from the instability and complexity of certain phenomena. Then the question becomes whether chance so defined can, or cannot, be deemed objective. In this connection, Poincaré remarks that what we observe now is the result of complex causes that since the beginning of the ages have been operating in the universe, always in the same direction, tending towards uniformity. It can be presumed that whatever today we call very small, or very complex, would have been perceived differently a very long time ago, or even in the very distant future, 'in millions and millions of centuries'. What we call 'chance' and attribute either to small variation in the initial conditions, or to complexity, is indeed relative, but

> it is not relative to this man or that, it is relative to the actual state of the world. It will change its meaning when the world becomes more uniform and all things are still more mixed. But then, no doubt, men will no longer be able to live, but will have to make way for other beings, [...] So that our criterion, remaining true for all men, retains an objective meaning (*ibid.*, p. 85).

The objectivity characterizing chance is therefore so with respect to man and the world he lives in. Of the man's world, chance is a fundamental component, which pervades not only physical phenomena, but all of human history. As Poincaré puts it,

> The greatest chance is the birth of a great man. It is only by chance that the meeting occurs of two genital cells of different sex that contain precisely, each on its side, the mysterious elements whose mutual reaction is destined to produce genius. It will be readily admitted that these elements must be rare, and that their meeting is still rarer. How little it would have taken to make the spermatozoid which carried them deviate from its course. It would have been enough to deflect it a hundredth part of an inch, and Napoleon would not have been born and the destinies of a continent would have been changed. No example can

give a better comprehension of the true character of chance (*ibid.*, p. 87).

On the contrary, Poincaré opposes the application of chance and probability to human action, and more generally to the moral sciences, and strongly criticizes Condorcet in this connection. The reason we tend to attribute human conduct to chance, he claims, lies with the complexity of causes underlying it. But such causes are not sufficiently complex to ground chance on them, because there are some invariant elements, something that is preserved in human conduct. With evident sarcasm, he remarks that

> When men are brought together, they no longer decide by chance and independently of each other, but react upon one another. Many causes come into action, they trouble the men and draw them this way and that, but there is one thing they cannot destroy, the habits they have of Panurge's sheep. And it is this that is preserved (*ibid.*, p. 88).

It is an open question whether Poincaré was a determinist or an indeterminist. In his *Dernières pensées*[26], he takes it as a matter of fact that science is determinist, and adds that, 'wrightly or wrongly', when investigating natural phenomena, scientists take determinism for granted. But he does not take sides on whether he believes this attitude is right or wrong. What is certain, is that Poincaré's account of chance is fully compatible with determinism – but also with indeterminism, for that matter.

As already observed, Poincaré's thoughts about chance impressed von Mises, who explicitly opened the door to indeterminism, and therefore to the existence of chance phenomena in nature. However, von Mises' indeterminism does not seem to imply any overarching metaphysical hypothesis about the world; it looks rather like an epistemological attitude, meant to accommodate within a unified framework all scientific knowledge.

On the contrary, indeterminism and the conviction that there is chance in nature are openly embraced by most upholders of the propensity theory of probability, starting with Peirce and Popper.

A balanced position with respect to the dichotomy determinism/indeterminism, inspired by Poincaré's work, is taken by Patrick Suppes. Starting from the tenet that also deterministic processes, if characterized by great complexity and/or instability, are unpredictable and can therefore be qualified as fortuitous, Suppes suggests that random sequences should be seen as 'the

---

[26] See Poincaré (1912b, English edition 1963).

limiting case of increasingly complex deterministic sequences', to which he adds that 'randomness is just a feature of the most complex deterministic sequences' (Suppes 1991, 1993a, p. 255). The traditional opposition between determinism, predictability and causality, on the one hand, and indeterminism, unpredictability and chance, on the other, is in this way superseded in favour of a view that accommodates determinism and indeterminism within a unique perspective. At the core of it we find the claim that neither determinism or indeterminism have an empirical foundation, so that the choice between them transcends experience[27].

---

[27] Specific arguments in favour of this thesis are to be found in Suppes (1993b). See also Suppes (1984) for his view of determinism and indeterminism. For an excellent overview of the doctrine of determinism see Earman (1986).

# 6

## The logical interpretation

### 6.1 Beginnings

The logical interpretation of probability can be seen as a natural development of the idea that probability is an epistemic notion, pertaining to our knowledge of facts, rather than to facts themselves. With respect to Laplace's classical interpretation, this approach stresses the logical aspect of probability, and regards the theory of probability as part of logic. According to Ian Hacking, such an attitude can be traced back to Leibniz. In fact,

> Leibniz was the first to insist that probability theory can serve in a branch of logic comparable to the theory of deduction. He was the first to try to axiomatize probability as an internal science. He saw how a generalized theory of games should be the foundation for making any quantitative decision in situations where one must act on inconclusive evidence. Far ahead of his time he thought that probabilities are relational and he insisted that judgments of probability are relative to the data available (Hacking 1971a, p. 597).

Leibniz is therefore seen as anticipating Carnap's programme of inductive logic, by conceiving a new kind of logic, aimed at serving to determine probabilities from the available data[1].

The idea that probability represents a sort of degree of certainty, more precisely the degree to which a hypothesis is supported by a given amount of information, was later worked out in some detail by the mathematician and logician Bernard Bolzano. Born in Prague in 1781, Bolzano was the son of a

---

[1] See also Hacking (1971b) and (1975).

German mother and an Italian father. After graduating in theology and being ordained a priest, in 1805 he was appointed professor of philosophy at the Charles-Ferdinand University in Prague. He hold that position until 1820, when he was dismissed from his chair as an activist of the 'Bohemian Enlightenment', a movement opposed to discrimination against the Czech Bohemians and fighting for social reforms. He then retreated to a village in Southern Bohemia, where he wrote some of his major works, including the treatise *Wissenschaftslehre* (1837), containing the most extensive exposition of his epistemology, considered to anticipate contemporary analytical philosophy[2]. In 1841 Bolzano went back to Prague, where he delivered many papers at the Bohemian Academy of Sciences, and died in 1848. Bolzano left an outstanding contribution to mathematics, partly contained in *Paradoxien des Unendlichen*, published posthumously in 1851.

Bolzano explicitly defines probability as a logical function holding between propositions. Broadly speaking, probability is seen as a 'degree of validity' (*Grad der Gültigkeit*) relative to a proposition expressing a hypothesis, with respect to other propositions, expressing the possibilities open to it. It is not possible to enter into the details of Bolzano's approach, based on the analysis of what he calls the 'variants' of propositions. It is, however, noteworthy that according to Bolzano probability is an objective notion, exactly like truth, from which probability derives. The main ingredients of logicism, namely the idea that probability is a logical relation between propositions, endowed with an objective character, are present in Bolzano's conception, which can be seen as a direct ancestor of Carnap's theory of probability as 'partial implication'.

## 6.2 The nineteenth century English logicists: De Morgan, Boole, Jevons

### Augustus De Morgan

Augustus De Morgan was born in India in 1806, son of a colonel of the Indian Army, and died in London in 1871. A Cambridge graduate, and later professor of mathematics at the University College, London, he was one of the most distinguished mathematicians of the nineteenth century and made substantial contributions to algebra, differential and integral calculus, and above all

---

[2] See for example Dummett (1993).

symbolic logic. His work had a great influence on Boole, who explicitly acknowledged it. He was also a libertarian, who did not hesitate to resign his post when he judged the council of the University College guilty of religious intolerance. After De Morgan's death, his wife Sophia Elizabeth described his life in a long *Memoir*, published in 1882, which also contains some of his correspondence.

De Morgan's major work in logic is the treatise *Formal Logic: or, The Calculus of Inference, Necessary and Probable* (1847). In this book, he claims that 'by degree of probability we really mean, or ought to mean, degree of belief' (De Morgan 1847, 1926, p. 198). While holding this position, he strongly opposes any attempt at defining probability as an objective feature of objects, like their physical properties. His attitude in this connection is radical:

> I throw away objective *probability* altogether, and consider the word as meaning the state of the mind with respect to an assertion, a coming event, or any other matter on which absolute knowledge does not exist (*ibid.*, p. 199).

However, De Morgan points out that he does not refer to actual belief, entertained by single persons, but rather to the kind of belief that a rational agent should adopt when evaluating probability. Therefore, to say that the probability of a certain event is three to one should be taken to mean 'that in the universal opinion of those who examine the subject, the state of mind to which a person *ought* to be able to bring himself is to look three times as confidently upon the arrival as upon the non-arrival' (*ibid.*, p. 200).

De Morgan also wrote various essays specifically devoted to probability, among which *Theory of Probabilities* (1837), and *An Essay on Probabilities, and on their Applications to Life, Contingencies and Insurance Offices* (1838). In the latter, he clarifies that 'the quantities which we propose to compare are the forces of the different impressions produced by different circumstances' (De Morgan 1838, p. 6), to which he adds that 'probability is the feeling of the mind, not the inherent property of a set of circumstances' (*ibid.*, p. 7). The terminology adopted by De Morgan and his insistence on terms such as 'degree of belief' and 'state of mind' in connection with probability, seem to bring him close to subjectivism. But unlike modern subjectivists like Bruno de Finetti, he refers to human mind as transcending individuals, not to the minds of single agents who evaluate probabilities.

## George Boole

George Boole was born in 1815 in Lincoln. Though to a certain extent self-taught, he became one of the most distinguished mathematicians and logicians of his time. Appointed professor of mathematics at Cork in 1849, he received many honours, and by the time of his death in 1864 was a member of a number of prestigious scientific institutions, including the Royal Society. Though his name is usually associated with Boolean algebra, he also made important contributions to other fields, like differential and integral calculus and, last but not least, probability. Among his works, the most famous is *An Investigation of the Laws of Thought, on Which are Founded the Mathematical Theories of Logic and Probabilities* (1854). A number of Boole's articles on logic and probability are collected in the volume *Studies in Logic and Probability* (1952), containing also a sketch of his life by Robert Harley. According to his biographer Desmond MacHale, Boole developed his theory of probability 'in the period 1851-1854, in a series of papers written in Cork and while on holiday in Lincoln'. MacHale adds that 'he was helped and greatly encouraged by W.F. Donkin, Savilian Professor of Astronomy in Oxford, who had himself written some important papers on the subject of probability. Boole was gratified that Donkin agreed with his results' (MacHale 1985, p. 215). It is not surprising that Boole's ideas interested William Donkin, who shared with Boole an epistemic view of probability, though Donkin, as argued in the next chapter, is closer to the subjective outlook.

Boole stresses that probability always makes reference to the state of knowledge, or to the information available to those who evaluate it. On the basis of such information, probability gives grounds for expectation. In Boole's words:

> *probability*, in its mathematical acceptation, has reference to the state of our knowledge of the circumstances under which an event may happen or fail. With the degree of information which we possess concerning the circumstances of an event, the reason we have to think that it will occur, or, to use a single term, our *expectation* of it, will vary. Probability is expectation founded upon partial knowledge (Boole 1854a, 1916, p. 258).

However, the claim that probability gives grounds for expectation should not be taken to mean that probability *is* degree of expectation. Boole is careful to avoid any such misinterpretation of his view and points out that

The rules which we employ in life-assurance, and in the other statistical applications of the theory of probabilities, are altogether independent of the *mental* phaenomena of expectation. They are founded on the assumption that the future will bear a resemblance to the past; that under the same circumstances the same event will tend to recur with a definite numerical frequency; not upon any attempt to submit to calculation the strength of human hopes and fears (*ibid.*, pp. 258-9).

In the same vein, in 'On a General Method in the Theory of Probabilities', Boole writes: 'probability I conceive to be not so much expectation, as a rational ground for expectation' (Boole 1854b, 1952, p. 292). The accent on rationality, together with the normative attitude towards the theory of probability, features a peculiar trait of the logical interpretation, marking a major difference from subjectivism. Within Boole's perspective, the normative character of probability derives from that of logic, to which it belongs. The 'laws of thought' are not meant to describe how the mind works, but rather how it should work in order to be rational[3]. This view is clearly expressed in *The Laws of Thought*, where Boole maintains that 'the mathematical laws of reasoning are, properly speaking, the laws of *right* reasoning only' (Boole 1854a, 1916, p. 428).

Boole embraces a logical perspective, according to which probability does not represent a relationships between events, but rather between propositions describing events. In Boole's words:

Although the immediate business of the theory of probability is with the frequency of the occurrence of events, and although it therefore borrows some of its elements from the science of number, yet as the expression of the occurrence of those events, and also of their relations, of whatever kind, which connect them, is the office of language, the common instrument of reason, so the theory of probabilities must bear some definite relation to logic. The events of which it takes account are expressed by propositions; their relations are involved in the relations of propositions. Regarded in this light, the object of the theory of probabilities may be thus stated: – Given the separate probabilities of any propositions to find the probability of another proposition. By the probability of a proposition, I here mean [...] the probability that in any

---

[3] Boole is usually said to combine a normative attitude towards logic with some sort of psychologism. On this point see William Kneale (1948) and the 'Introduction' (Part I by Ivor Grattan-Guinness and Part II by Gérard Bornet) in Boole (1997).

particular instance, arbitrarily chosen, the event or condition which it affirms will come to pass (Boole 1851, 1952, pp. 250-1).

Accordingly, the theory of probability becomes 'coextensive with that of logic, and [...] it recognizes no relations among events but such as are capable of being expressed by propositions' (*ibidem*).

Boole distinguishes between two kinds of objects of interest to probability. The first is games of chance, the second relates to observable phenomena belonging to the natural and social sciences. Games of chance confront us with a peculiar kind of problems, where the ascertainment of data is in itself a way of measuring probabilities. Events of this kind are called *simple*. Sometimes such events are combined to form a *compound* event, as when it is asked what is the probability of obtaining twice a six in two successive throws of a die. Instead, when dealing with the phenomena encountered in nature we are bound to appeal to frequencies in order to measure probabilities, and we face compound events. Simple events, described by simple propositions, and compound events, described by compounded propositions, are the objects of he theory of probability:

> the probabilities of events, or of combinations of events, whether deduced from a knowledge of the peculiar constitution of things under which they happen, or derived from the long-continued observation of past series of their occurrences and failures, constitute, in all cases, our data (Boole 1854a, 1916, p. 260).

Simple propositions are combined to form compounded propositions by means of the logical relations of conjunction and disjunction, and the dependence of the occurrence of certain events upon others can be represented by conditional propositions. Thus, the methods of logic provide an apt tool for using propositions to represent all of the events subject to probability. Having so argued, Boole defines the measure of probability of an event as 'the ratio of the number of cases favourable to that event, to the total number of cases favourable or contrary, and all equally possible' (*ibid.*, p. 258), a definition that he claims to borrow from Poisson. Based on this definition, the fundamental rules for calculating compounded probabilities are presented by Boole in such a way, as to show their intimate relation with logic, and more precisely with his algebra. The conclusion attained is that there is a 'natural bearing and dependence' (*ibid.*, p. 287) between the numerical measure of probability and the algebraic representation of the values of logical expressions.

Though, as we have seen, the idea of replacing events by propositions does not start with Boole, being clearly present in the earlier work of Bolzano, 'only with Boole's construction of a logical calculus of propositions did such a replacement have any material consequence' (Hailperin 1976, p. 132). In fact, Boole puts that idea at the basis of a programme, aiming

> to obtain a general method by which, given the probabilities of any events whatsoever, be they simple or compound, dependent or independent, conditioned or not, one can find the probability of some other event connected with them, the connection being either expressed by, or implicit in, a set of data given by logical equations (Boole, 1854a, 1916, p. 287).

In other words, Boole sets forth the logicist programme, to be resumed by Carnap about a hundred years later.

An important aspect of Boole's perspective is his criticism of Laplace's rule of succession, more particularly of the principle of insufficient reason underpinning the rule. So goes Boole's text:

> It is been said, that the principle involved [in the rule of succession] [...] is that of the equal distribution of our knowledge, or rather of our ignorance – the assigning to different states of things of which we know nothing, and upon the very ground that we know nothing, equal degrees of probability. I apprehend, however, that this is an arbitrary method of procedure (*ibid.*, p. 386).

When discussing the probability of judgments, in Chapter XXI of *The Laws of Thought*, Boole enforces his criticism, saying that 'adopting the methods of Laplace and Poisson [...] we are relying *wholly* upon a doubtful hypothesis, – the independence of individual judgments' (*ibid.*, p. 410). However, a detailed analysis of Boole's tentative solution to these problems falls beyond the scope of the present account[4].

## William Stanley Jevons

The economist and logician William Stanley Jevons was born in Liverpool in 1835 and died in Hastings in 1882. He was educated in various scientific disciplines, including mathematics, biology, chemistry and geology. After a

---

[4] The reader is referred to Hailperin (1976) for a technical account of Boole's theory of probability.

period in Australia as an assayer of the Sydney Mint, Jevons went back to England and graduated in philosophy and education at University College London. He was professor of logic and moral philosophy at Owens College in Manchester, and then professor of political economy at University College London from 1876 to 1881. In a sketch of Jevons' life and work, John Maynard Keynes describes him as 'a good Victorian', grown up in a family of 'educated nonconformists' (Keynes, 1936, 1972, p. 111), a man of culture, deeply concerned with social and economical problems. Like Francis Galton, Jevons strongly believed that the analysis of social phenomena could benefit from the adoption of quantitative methods, a conviction not shared by Keynes.

As to probability, we read in Jevons' major epistemological work, namely *The Principles of Science* (1873), that *'probability belongs wholly to the mind'* (Jevons 1873, 1877, p. 198). Although embracing an epistemic approach to probability, Jevons opposes the adoption of the notion of 'degree of belief' to define probability, because he finds it ambiguous. Against both Donkin and his teacher at University College Augustus De Morgan, who use this terminology, he maintains that 'the nature of *belief* is not more clear [...] than the notion which it is used to define. But an all-sufficient objection is, that *the theory does not measure what the belief is, but what it ought to be'* (*ibid.*, p. 199). In view of this, Jevons prefers 'to dispense altogether with this obscure word belief, and to say that the theory of probability deals with *quantity of knowledge'* (*ibidem*). If defined in this way, probability can provide a suitable guide of belief and action. In his words: 'the value of the theory consists in correcting and guiding our belief, and rendering one's states of mind and consequent actions harmonious with our knowledge of exterior conditions' (*ibidem*).

Jevons shows great confidence in the utility and power of probability. He has the merit of establishing a close link between probability and induction. In this regard, he declares he is convinced 'that it is impossible to expound the methods of induction in a sound manner, without resting them upon the theory of probability' (*ibid.*, p. 197). Moreover, he holds that

> No inductive conclusions are more than probable, and [...] the theory of probability is an essential part of logical method, so that the logical value of every inductive result must be determined consciously or unconsciously, according to the principle of the inverse method of probability (*ibid.*, p. xxix).

These claims notwithstanding, Jevons way of dealing with induction is not considered particularly innovative[5].

A controversial aspect of Jevons' approach to probability is his defence of Laplace's rule against the criticism of Boole. In this connection, he says that although

> The English writers Bayes and Price are [...] undoubtedly the first who put forward any distinct rules on the subject [...] it was reserved to the immortal Laplace to bring to the subject the full power of its genius, and carry the solution of the problem almost to perfection (*ibid.*, p. 261).

Jevons grants to Laplace's critics that the method he developed, especially his principle of insufficient reason, is to a certain extent arbitrary, but adds that it is nevertheless a useful method, of great help in situations characterized by lack of knowledge. In Jevons' words:

> It must be allowed that the hypothesis adopted by Laplace is in some degree arbitrary, so that there was some opening for the doubt which Boole has cast upon it. [...] But it may be replied [...] that the supposition of an infinite number of balls treated in the manner of Laplace is less arbitrary and more comprehensive than any other that can be suggested (*ibid.*, p. 256).

His conclusion is that Laplace's solution

> is only to be accepted in the absence of all better means, but like other results of the calculus of probability, it comes to our aid when knowledge is at an end and ignorance begins, and it prevents us from over-estimating the knowledge we possess (*ibid.*, p. 269).

When reading Jevons, one is impressed by his deeply probabilistic conception of science, which results in his pragmatical view. Statements like: 'the certainty of our scientific inferences [is] to a great extent a delusion' (*ibid.*, p. xxxi), or 'the truth or untruth of a natural law, when carefully investigated, resolves itself into a high or low degree of probability' (*ibid.*, p. 217) show that he regarded human knowledge as essentially based on probability. To his eyes, knowledge is of necessity incomplete: 'in the writings of some recent philosophers, especially of Auguste Comte, and in some degree John Stuart Mill, there is an erroneous tendency to represent our knowledge as assuming an

---

[5] This is the conclusion reached in Laudan (1973), where Jevons' view of induction is tackled in some detail.

approximately complete character' (*ibid.*, p. 752). In addition, science is based on the assumption of the uniformity of nature, as to which he maintains that 'those who so frequently use the expression Uniformity of Nature seem to forget that the Universe might exist consistently with the laws of nature in the most different conditions' (*ibid.*, p. 749). Under such conditions, we have to appeal to probability. This involves severe limitations, because probability does not tell us much with regard to what happens in the short run, but represents the best tool we have for facing the future:

> All that the calculus of probability pretends to give, is *the result in the long run*, as it is called, and this really means in *an infinity of cases*. During any finite experience, however long, chances may be against us. Nevertheless the theory is the best guide we can have (*ibid.*, p. 261).

## 6.3 John Maynard Keynes

### Probability as a logical relation

John Maynard Keynes (1883-1946) is perhaps the most famous economist of the twentieth century, and the author of overly influential books, like *A Treatise on Money* (1930) and *General Theory of Employment, Interest and Money* (1936). Besides playing a crucial role in public life as a political advisor, he was a supporter of the arts and a successful writer. Son of the logician John Neville Keynes, Maynard was educated at Eton and Cambridge, where he was later scholar and member of King's College. There he was a most active member of cultural life. Among other things, he took part in the 'Apostles' meetings and the 'Bloomsbury group', and intensively interacted with outstanding philosophers, like George Edward Moore, Bertrand Russell, John McTaggart and Frank Ramsey. His contribution to probability is contained in the volume *A Treatise on Probability*.

According to his biographer Roy Forbes Harrod, Keynes worked on the *Treatise* in the years 1906-1911[6]. Though by that time the book was all but completed, Keynes could not prompt its final revision until 1920, due to his many political commitments. When it finally appeared in print in 1921, the book was very well received, partly because of its author's fame as an

---

[6] See Harrod (1951). On Keynes' life and work see also Skidelsky (1983-1992).

economist and political adviser, partly because it was the first systematic work on probability by an English writer after Venn's *The Logic of Chance*, published forty-five years earlier. A review of the *Treatise* by Charlie Dunbar Broad opens with this passage: 'Mr Keynes's long awaited work on Probability is now published, and will at once take its place as the best treatise on the logical foundations of the subject' (Broad 1922, p. 72), and closes as follows: 'I can only conclude by congratulating Mr Keynes on finding time, amidst so many public duties, to complete this book, and the philosophical public on getting the best work on Probability that they are likely to see in this generation' (*ibid.*, p. 85)[7].

In the portrait 'Keynes as a Philosopher', included in the collection *Essays on John Maynard Keynes*, edited by his nephew Milo Keynes, Richard Bevan Braithwaite writes that

> The *Treatise* was enthusiastically received by philosophers in the empiricist tradition. [...] The welcome given to Keynes's book was largely due to the fact that his doctrine of probability filled an obvious gap in the empiricist theory of knowledge. Empiricists had divided knowledge into that which is 'intuitive' and that which is 'derivative' (to use Russell's terms), and he regarded the latter as being passed upon the former by virtue of there being a logical relationships between them. Keynes extended the notion of logical relation to include probability relations, which enabled a similar account to be given of how intuitive knowledge could form the basis for rational belief which fell short of knowledge (Braithwaite 1975, pp. 237-8).

Braithwaite is certainly right in pointing out that the main trend of empiricist philosophy at the time when Keynes' *Treatise* was published was oriented more towards the deductive aspects of science than to probability – suffice it to think of Bertrand Russell. However, one should not forget that, as we have seen, the logical approach to probability already counted a number of supporters in England. Besides, a similar approach was being adopted in the same years by Ludwig Wittgenstein and Friedrich Waismann, and in the same Cambridge where Keynes and Russell were moving their steps another influential figure, William Ernest Johnson, was taking the same path. In the 'Preface' to his book Keynes recognizes his debt to Johnson, and more

---

7. Regarding this statement, Braithwaite in his obituary of Keynes remarks that 'Broad's prophecy has proved correct' (Braithwaite 1946, p. 284).

generally to the Cambridge philosophical setting, seen as an ideal continuation of the great empiricist tradition 'of Locke and Berkeley and Hume, of Mill and Sidgwick, who, in spite of their divergencies of doctrine, are united in a preference for what is matter of fact, and have conceived their subject as a branch rather of science than of creative imagination' (Keynes 1921, pp. v-vi).

The theory of probability is conceived by Keynes as a branch of logic, more precisely as that part of logic which deals with arguments that are not conclusive, but can be said to have a greater or less degree of inconclusiveness. 'Part of our knowledge we obtain direct; and part by argument. The Theory of Probability is concerned with that part which we obtain by argument, and treats of the different degrees in which the results so obtained are conclusive or inconclusive' (*ibid.*, p. 3). Like the logic of conclusive arguments, the logic of probability investigates the general principles of inconclusive arguments. Both certainty and probability depend on the amount of knowledge that the premises of an argument convey to support the conclusion, the difference being that certainty obtains when the amount of available knowledge authorizes full belief, while in all other cases one obtains degrees of belief. Certainty can therefore be seen as the limiting case of probability. Keynes stresses the relational character of the notion of belief:

> The terms *certain* and *probable* – he says – describe the various degrees of rational belief about a proposition which different amounts of knowledge authorize to entertain. All propositions are true or false, but the knowledge we have of them depends on our circumstances [...] it is without significance to call a proposition probable unless we specify the knowledge to which we are relating it (*ibidem*).

Such knowledge is described by a set of propositions, constituting the premises of an argument, which stand in a logical relationship with the conclusion, describing a hypothesis of some sort. Probability resides in this logical relationship, and its value will vary according to the information conveyed by the premises of the arguments involved:

> As our knowledge or our hypothesis changes, our conclusions have new probabilities, not in themselves, but relatively to these new premises. New logical relations have now become important, namely those between the conclusions which we are investigating and our new assumptions; but the old relations between the conclusions and the former assumptions still exist and are just as real as these new ones (*ibid.*, p. 7).

On this basis, Keynes develops a theory of comparative probability, in which conditional probabilities are ordered in terms of a relation of 'more' or 'less probable', and combined into compound probabilities.

## Rationality and the role of intuition

The logical character of probability goes hand in hand with its *rational* character. Like his predecessors – Boole, for instance – Keynes wants to develop a theory of the reasonableness of degrees of belief on logical grounds. According to him, the theory of probability as a logical relation: 'is concerned with the degree of belief which it is *rational* to entertain in given conditions, and not merely with the actual beliefs of particular individuals, which may or may not be rational' (*ibid.*, p. 4). Keynes' logical interpretation gives the theory of probability a normative value: 'we assert that we *ought* on the evidence to prefer such and such a belief. We claim rational grounds for assertions which are not conclusively demonstrated' (*ibid.*, p. 5). Probability so conceived is *objective*, and its objectivity is warranted by its logical character:

> What particular propositions we select as the premises of *our* argument naturally depends on subjective factors peculiar to ourselves; but the relations, in which other propositions stand to these, and which entitle us to probable beliefs, are objective and logical (*ibid.*, p. 4).

In other words, the logical relations between the premises and the conclusion of inconclusive arguments provide objective grounds for belief, and entitle the belief based on them to be qualified as *rational*. As to the character of the logical relations themselves, Keynes says that 'we cannot analyse the probability-relation in terms of simpler ideas' (*ibid.*, p. 8). They are therefore taken as primitive, and their justification is left to our intuition. The importance of intuition within Keynes' perspective can be associated with the influence exercised on him by George Edward Moore, another outstanding figure of the Cambridge milieu[8].

The approach adopted by Keynes can be described as a moderate form of logicism, quite different from the strictly formal approach later developed by Carnap. Keynes' logicism is pervaded by a deeply felt need not to lose sight of ordinary speech and practice, and assigns an important role to intuition and individual judgment. Moreover, Keynes is suspicious of a purely formal

---

[8] On this point see Gillies (2000), especially Chapter 3, which contains a useful discussion of Keynes' theory of probability. See also see Gillies (1988).

treatment of probability, and of the adoption of mechanical rules for the evaluation of probability. This attitude inspires a controversial feature of Keynes' theory, namely his claim that such relations are not always measurable, nor comparable:

> By saying that not all probabilities are measurable, I mean that it is not possible to say of every pair of conclusions, about which we have some knowledge, that the degree of our rational belief in one bears any numerical relation to the degree of our rational belief in the other; and by saying that not all probabilities are comparable in respect of more and less, I mean that it is not always possible to say that the degree of our rational belief in one conclusion is either equal to, greater than, or less than the degree of our belief in another (*ibid.*, p. 34).

In other words, Keynes admits of some probability relations which are intractable by the calculus of probabilities. Far from being worrying to him, this aspect shows Keynes' conviction that intuition plays an important role with respect to probability.

The measurement of probability rests on the principle of insufficient reason: 'In order that numerical measurement may be possible, we must be given a number of *equally* probable alternatives' (*ibid.*, p. 41). After having admitted that, Keynes severely criticizes Laplace's principle, and discusses at length the paradoxes it raises[9]. Keynes' main philosophical argument against Laplace's principle, which he calls 'principle of indifference' to stress the role of individual judgment in the ascription of equal probability to all possible alternatives, is that 'the rule that there must be no ground for preferring one alternative to another, involves, if it is to be a guiding rule at all, and not a *petitio principii*, an appeal to judgments of *irrelevance*' (*ibid.*, pp. 54-5). The judgment of indifference between various alternatives has to be substantiated with the assumption that there could be no further information, on account of which one might change such judgment itself. While in the case of games of chance an assumption of this kind can be made without problems, most cases encountered in everyday life are characterized by a complexity that makes such an assumption rather arbitrary. For Keynes, the extension of the principle of insufficient reason to cover all possible applications is the expression of a superficial way of addressing probability, regarded as a product of ignorance, rather than knowledge. On the contrary, Keynes maintains that the judgment of

---

[9] Also on this point, see Gillies (2000), Chapter 3.

indifference between the available alternatives should not be grounded on ignorance, but on knowledge. In fact, the application of the principle in question has to be preceded by an act of discrimination between relevant and irrelevant elements of the available information, and by the decision to neglect certain pieces of evidence. This can only be guided by knowledge, not by ignorance.

Keynes' distrust in the practice of unrestrictedly applying principles holding within a restricted range, regards not only the principle of insufficient reason, but extends also to the 'principle of induction', taken as a method of establishing empirical knowledge from the observation of a large number of cases. In particular, Keynes is suspicious of the inference of general principles on an inductive basis, including causal laws and the principle of uniformity of nature. He reaffirms the relational character of inductive inference: 'an inductive argument affirms, not that a certain matter of fact *is* so, but that *relative to certain evidence* there is a probability in its favour' (*ibid.*, p. 221). In addition, one should not forget that, given the logical character of the probability relation,

> The validity and reasonable nature of inductive generalization is [...] a question of logic and not of experience, of formal and not of material laws. The actual constitution of the phenomenal universe determines the character of our evidence; but it cannot determine what conclusions *given* evidence *rationally* supports (*ibidem*).

## Analogy, relevance and weight

On the same basis, Keynes criticizes inferential methods entirely grounded on repeated observations, like the calculation of frequencies. Against this attitude, he claims that the similarities and dissimilarities among events must be carefully considered before quantitative methods can be applied. In this connection, a crucial role is played by analogy, which becomes a prerequisite of statistical inductive methods based on frequencies. Keynes writes:

> To argue from the *mere* fact that a given event has occurred invariably in a thousand instances under observation, without any analysis of the circumstances accompanying the individual instances, that it is likely to occur invariably in future instances, is a feeble inductive argument, because it takes no account of the Analogy (*ibid.*, p. 407).

The insistence upon analogy is a central feature of Keynes' approach. In an attempt to provide a logical foundation for such a notion, Keynes appeals to the assumption that the variety encountered in the world has to be of a limited kind:

> As a logical foundation for Analogy, [...] we seem to need some such assumption as that the amount of variety in the universe is limited in such a way that there is no one object so complex that its qualities fall into an infinite number of independent groups (*i.e.* groups which might exist independently as well as in conjunction); or rather hat none of the objects about which we generalise are as complex as this; or at least that, though some objects may be infinitely complex, we sometimes have a finite probability that an object about which we seek to generalise is not infinitely complex (*ibid.*, p. 258).

This assumption confers a finitistic character to Keynes' approach, criticized, among others, by Carnap and Jeffreys. The principle of limited variety is attacked on a somewhat different basis by Ramsey[10], who claims to see 'no logical reason for believing any such hypotheses; they are not the sort of things of which we could be supposed to have a priori knowledge, for they are complicated generalizations about the world which evidently may not be true' (Ramsey 1991a, p. 297).

Another important aspect of Keynes' theory is connected with the notion of *weight* of arguments. Like probability, the weight of inductive arguments varies according to the amount of evidence. But while probability is affected by the proportion between favourable and unfavourable evidence, weight increases as relevant evidence – taken as the sum of positive and negative observations – increases. In Keynes' words:

> As the relevant evidence at our disposal increases, the magnitude of the probability of the argument may either decrease or increase, according as the new knowledge strengthens the unfavourable or the favourable evidence; but *something* seems to have increased in either case, – we have a more substantial basis upon which to rest our conclusion. I express this by saying that an accession of new evidence increases the *weight* of an argument. New evidence will sometimes decrease the probability of an argument, but it will always increase its 'weight' (Keynes 1921, p. 71).

---

[10] See especially 'On the Hypothesis of Limited Variety' and 'Induction: Keynes and Wittgenstein', both in Ramsey (1991a).

The concept of weight is strictly intertwined with that of relevance, because to say that a piece of evidence is relevant is the same as saying that it increases the weight of an argument. Therefore, Keynes' stress on weight backs the importance of the notion of relevance within his theory of probability.

Keynes also addresses the issue of whether the weight of arguments should be made to bear upon action choice. He writes: 'the question comes to this – if two probabilities are equal in degree, ought we, in choosing our course of action, to prefer that one which is based on a greater body of knowledge?' (*ibid.*, p. 313). This issue, he claims, has been neglected by the literature on action choice, essentially based on the notion of mathematical expectation. Having said this, Keynes adds: 'the question appears to me highly perplexing, and it is difficult to say much that is useful about it' (*ibidem*). The discussion of these topics leads to a sceptical conclusion, reflecting Keynes' distrust in a strictly mathematical treatment of the matter:

> The hope, which sustained many investigators in the course of the nineteenth century, of gradually bringing the moral sciences under the sway of mathematical reasoning, steadily recedes – if we mean, as they meant, by mathematics the introduction of precise numerical methods. The old assumptions, that all quantity is numerical and that all quantitative characteristics are additive, can be no longer sustained. Mathematical reasoning now appears as an aid in its symbolic rather that in its numerical character (*ibid.*, p. 316).

As already noticed, Keynes' caution is meant to leave room for individual judgment and intuition. It is worth noticing that his attitude in this connection is inextricably linked to the conviction that probability is objective, and not always measurable. As the next chapter will illustrate, this clutch of problems will be addressed by subjectivists in a rather different spirit.

According to Keynes, the validity of inductive arguments cannot be made to depend on their success, and it is not undermined by the fact that some events which have been predicted do not actually take place. Induction allows us to say that on the basis of a certain piece of evidence a certain conclusion is reasonable, not that it is true. Awareness of this fact should inspire a cautious attitude towards inductive predictions, and Keynes warns against the danger of making predictions obtained by detaching the conclusion of an inductive argument. This features a typical aspect of the logical interpretation of probability, that has been at the centre of a vast debate, in which Carnap also took part[11].

The insistence on the logical and non-empirical character of probability relations, urges Keynes to oppose any attempt at grounding probabilistic inference on success. This is stressed by Anna Carabelli, who writes that

> Keynes was [...] critical of the positivist *a posteriori* criterion of the validity of induction, by which the inductive generalization was valid as far as the prevision based on it will prove successful, that is, will be confirmed by subsequent facts. [...] On the contrary, the validity of inductive method, according to Keynes, did not depend on the success of its prediction, or on its empirical confirmation (Carabelli 1988, p. 66).

Unfortunately, Carabelli goes on to say that 'notably, that was what made the difference between Keynes's position and that of those later logico-empiricists, like R. Carnap, who analysed induction from what he called the "confirmation theory" point of view' (*ibidem*). This last claim is misleading, because Carnap's confirmation theory is not so closely linked to the criterion of success as Carabelli claims. As argued in Section 6.6, in his late writings Carnap appeals to 'inductive intuition' to justify induction, thereby embracing a position close to that of Keynes.

## Ramsey's criticism

It is worth closing this presentation of Keynes' perspective by recollecting that soon after the publication of the *Treatise*, Ramsey published a critical review, challenging some of its central issues, like the conviction that there are unknown probabilities, the principle of limited variety, and the very idea that probability is a logical relation[12]. Ramsey is indeed very critical of this idea, and objects to Keynes that 'there really do not seem to be any such things as the probability relations he describes' (Ramsey 1990a p. 57)[13]. After Ramsey's premature death in 1930, Keynes wrote an obituary, containing an explicit concession to Ramsey's criticism. There he says:

---

[11] See, for instance, Kyburg (1968) and the discussion following it, with comments by Y. Bar-Hillel, P. Suppes, K.R. Popper, W.C. Salmon, J. Hintikka, R. Carnap, H. Kyburg jr.

[12] See Ramsey (1922). Further comments are to be found in Ramsey's 'Truth and Probability', published in Ramsey (1931) and (1990a).

[13] See also Ramsey (1991a), p. 273, where we find the claim that: 'there are no such things as these relations'.

Ramsey argues, as against the view which I had put forward, that probability is concerned not with objective relations between propositions but (in some sense) with degrees of belief, and he succeeds in showing that the calculus of probabilities simply amounts to a set of rules for ensuring that the system of degrees of belief which we hold shall be a *consistent* system. Thus the calculus of probabilities belongs to formal logic. But the basis of our degrees of belief – or the *a priori* probabilities, as they used to be called – is part of our human outfit, perhaps given us merely by natural selection, analogous to our perceptions and our memories rather than to formal logic. So far I yield to Ramsey – I think he is right (Keynes 1930, 1972, p. 339).

These claims have long been debated among those who believe that Keynes embraced the subjectivist interpretation of probability after Ramsey's criticism, and those who are instead convinced that Keynes never changed his mind in a substantial way[14].

However, what Keynes says in the continuation of the above quoted passage seems eloquent:

> But in attempting to distinguish 'rational' degrees of belief from belief in general he [Ramsey] was not yet, I think, quite successful. It is not getting to the bottom of the principle of induction merely to say that it is a useful mental habit (*ibidem*).

Keynes' insistence on separating rational belief from actual belief sides him with logicism, as opposed to subjectivism.

## 6.4 William Ernest Johnson

William Ernest Johnson was born in Cambridge in 1858 to a Baptist family, professing liberal ideas in politics, religion and society. He received his first education from his father, who was master of the Llandaff House school in Cambridge. Severe health problems, that started when he contracted asthma at the age of eight, affected him throughout his life. He studied mathematics and moral science, and became lecturer in the University of Cambridge, first on the

---

[14] Anna Carabelli, for instance, believes that 'Keynes did not change substantially his view on probability' (Carabelli 1988, p. 255). For a discussion of this topic see Gillies (2000).

mathematical theory of economics, and later on moral science and logic. When he was elected Fellow of King's College in 1902, he was a most respected philosopher and intellectual. By the time of his death in 1931, he had been elected member of the British Academy and honoured by degrees from the Universities of Manchester and Aberdeen. He is buried in the cemetery of Grantchester, near the Orchard tea gardens, where many Cambridge intellectuals used to convene at Johnson's time, and which is still a visitor's attraction.

In an obituary of Johnson, Keynes describes him as 'one of the acutest philosophers in Cambridge', and adds that 'he had an immense influence through his love of discussion and conversation on almost all Cambridge moral scientists of the last 40 years, a great number of whom were his pupils' (Keynes 1931, 1972 p. 349). A memoir written by Broad for the British Academy depicts Johnson as a man of deep culture, versed in literature and music, naturally inclined to teaching, and above all a strong believer in the powers of reason, a son of that 'generation of Radical Nonconformists' that 'really believed in reason and reasonableness, and strove according to its lights to apply them to the solution of philosophic, political, and international problems' (Broad 1931, p. 513).

Partly because of his poor health, partly because of various difficulties he had to face in his life, and partly because of the high standard of achievement he set himself, Johnson did not publish much. His most important philosophical work is *Logic*, published in three volumes between 1921 and 1924. As to this book, Broad observes that it 'is much more than a treatise on deductive and inductive logic as ordinarily understood. It contains most valuable and original chapters on fundamental problems of epistemology, metaphysics, and even psychology' (*ibid.*, p. 507).

Johnson adopts a 'philosophical' approach to logic, taken as 'the analysis and criticism of thought' (Johnson 1921, 1964, p. xiii). His first concern in dealing with the discipline of logic is to distinguish between 'the epistemic aspect of thought', connected with 'the variable conditions and capacities for its acquisition', and its 'constitutive aspect', referring to 'the content of knowledge which has in itself a logically analysable form' (*ibid.*, pp. xxxiii-iv). The epistemic and grammatical aspects of logic, regarded as separate but strictly intertwined, are the two poles of Johnson's treatment of the matter. Johnson's stress on the epistemic aspects of logic sets him somehow apart from the period's main tendency towards formal logic. In a sympathetic spirit, Keynes points out that Johnson was the first to exercise the epistemic side of logic, the links between logic and the psychology of thought. In a school of thought whose natural leanings were towards formal logic, he was exceptionally well equipped to write formal logic himself and to criticize everything which was being contributed to the subject along formal lines (Keynes 1931, 1972, p. 349).

Commenting on the same issue, Broad observes that

> in matters of pure logic Johnson remained completely unaffected by the work of Dr. Wittgenstein or of recent German theorists on the foundations of mathematics, such as Weyl, Hilbert, & c. Whatever may be the outcome of these later developments much of Johnson's work will remain untouched, and we shall often return with relief and profit to his solid English sanity from the wilder flights of Teutonic speculation; but it may well be that his treatment of the laws of thought and the foundations of mathematics will become out of date (Broad 1931, p. 509).

As a matter of fact, Johnson's work is widely neglected by the contemporary debate on epistemology, though it contains a great many elements of originality.

Johnson's *Logic* was planned to include a fourth book, dealing with probability. Of this book, which was never completed, Johnson wrote the first three chapters, which appeared posthumously in *Mind* in 1932. Also dedicated to probability is the 'Appendix on Eduction' closing the third volume of *Logic*. Johnson embraces a logical notion of probability, according to which probability is to be attached to propositions. He then opposes any definition of probability as a property of events: 'familiarly we speak of the probability of an event; but, in my view, such an expression is not justifiable' (Johnson 1932, p. 2). Instead, for Johnson

> Probability is a character, variable in quantity or degree, which may be predicated of a proposition considered in its relation to some other proposition. The proposition to which the probability is assigned is called the proposal, and the proposition to which the probability of the proposal refers is called the supposal (*ibid.*, p. 8).

Here *proposal* and *supposal* represent what is usually called hypothesis and evidence. As Johnson puts it, a peculiar feature of the theory of probability is that when dealing with it 'we have to recognise not only the two assertive attitudes of acceptance and rejection of a given assertum, but also a third attitude, in which judgment as to its truth or falsity is suspended; and [...] probability can only be expounded by reference to such an attitude towards a given assertum' (*ibid.*, p. 2). If the act of suspending judgment is a mental fact, and as such is the competence of psychology, the treatment of probability taken in reference to such act is also strongly connected to logic, because logic provides the norms to be imposed on it. In Johnson's words:

> The logical treatment of probability is related to the psychological treatment of suspense of judgment in the same way as the logical treatment of the proposition is related to the psychological treatment of belief. Just as logic lays down some conditions for correct belief, so also it lays down conditions for correcting the attitude of suspense of judgment. In both cases we hold that logic is normative, in the sense that it imposes imperatives which have significance only in relation to presumed errors in the processes of thinking: thus, if there are criteria of truth, it is because belief sometimes errs. Similarly, if there are principles for the measurement of probability, it is because the attitude of suspense towards an assertum involves a mental measurable element, which is capable of correction. We therefore put forward the view, that probability is quantitative because there is a quantitative element in the attitude of suspense of judgment (*ibid.*, pp. 2-3).

The above passage describes very neatly in what sense for Johnson probability falls within the realm of logic.

Johnson's logical account of probability starts with a distinction between three types of probability statements according to their form. They are:

> (1) The singular proposition, *e.g.*, that the next throw will be heads, or that this applicant for insurance will die within a year; (2) The class-fractional proposition, *e.g.*, that, of the applicants to an insurance office, 3/4 of consumptives will die within a year; or that 1/2 of a large number of throws will be heads; (3) The universal proposition, *e.g.*, that all men die before the age of 150 years (*ibid.*, p. 2).

According to Johnson, these three types of probability statements are only superficially similar, but give rise to different problems, and therefore should not be confused. In a more widely used terminology, Johnson's worry is to distinguish between propositions referring to (1) a generic individual randomly chosen from a population, (2) a finite sample or population, (3) an infinite population. The distinction appears crucial for both understanding and evaluating statistical inference, and Johnson has the merit of having pointed it out[15].

Johnson's reason for introducing the above distinction lies with his conviction that probability, conceived as the relation between proposal and supposal, presents two distinct aspects: *constructional* and *inferential*. Now, in conceiving the constructional relation between any two given propositions, we have to note [...] both the form of each proposition taken by itself, and the process by which one proposition is constructed from the other' (*ibid.*, p. 4). In the case of probability, the form of the propositions involved and the way in which the proposal is constructed by modification of the supposal, will determine the constructional relation between them. On such constructional relation is in turn based the inferential relation, 'namely, the measure of probability that should be assigned to the proposal as based upon assurance with respect to the truth of the supposal' (*ibidem*). A couple of examples, taken from Johnson's exposition, will illustrate his position:

> Let the proposal be that 'The next throw of a certain coin will give heads'. Let the supposal be that 'the next throw of the coin will give heads or tails'. Then the relation of probability in which the proposal stands to the supposal is determined by the relation of the predication 'heads' to the predication 'heads or tails'. Or. To take another example, let the proposal be that 'the next man we meet will be tall and red-haired', and the supposal that 'the next man we meet will be tall'. Then the relation of predication 'tall and red-haired' to the predication 'tall' will determine the probability to be assigned to the proposal as depending on the supposal. These two cases illustrate the way in which the logical conjunctions 'or' and 'and' enter into the calculus of probability (*ibid.*, p. 8).

Along these lines, Johnson builds a rather sophisticated logical theory of probability. The relation of partial implication between propositions, on which this is ultimately based, bears a strict resemblance to the theory worked out some years later by Carnap, with the difference that Carnap adopted an improved definition of the 'content' of a proposition, making use of formal semantics.

---

[15] Some remarks on the relevance of the distinction made by Johnson are to be found in Costantini and Galavotti (1987).

Johnson's greatest merit is no doubt that of having devised the probabilistic property of exchangeability, introduced under the name of *Permutation postulate* in the 'Appendix on Eduction' published in the third volume of *Logic*. According to such property, probability is invariant with respect to permutation of individuals, to the effect that exchangeable probability functions assign probability in a way that depends on the number of experienced cases, irrespective of the order in which they have been observed. As it will be argued in what follows, this property plays a crucial role within Carnap's inductive logic – where it is named 'symmetry' – and de Finetti's subjective theory of probability – where it is called by the more widespread term 'exchangeability'. Johnson's discovery of this result left some trace in Ramsey's work, but remained almost ignored until the statistician John Irving Good called attention to it in his monograph *The Estimation of Probabilities. An Essay on Modern Bayesian Methods*, which opens with the following words: 'This monograph is dedicated to William Ernest Johnson, the teacher of John Maynard Keynes and Harold Jeffreys' (Good 1965, p. v). Good made extensive use of Johnson's ideas in a Bayesian framework[16].

The insight of Johnson's treatment of probability was not fully grasped by his contemporaries. Suffice it to mention Broad's comment on Johnson's 'Appendix on Eduction': 'about the Appendix all I can do is, with the utmost respect to Mr Johnson, to parody Mr Hobbes's remark about the treatises of Milton and Salmasius: "very good mathematics; I have rarely seen better. And very bad probability; I have rarely seen worse"' (Broad 1924, p. 379). But it should be added that Johnson's ideas certainly influenced Keynes, Ramsey, Jeffreys, and many others.

## 6.5 Viennese logicism: Wittgenstein and Waismann

### Ludwig Wittgenstein

Ludwig Wittgenstein is widely known as one of the most outstanding figures on the nineteenth century philosophical scene. Born in 1889 in Vienna to a very wealthy family that occupied a prominent position in the social and artistic life of the Austrian capital, Wittgenstein received a general education,

---

[16] See Zabell (1982) for a historical and theoretical discussion of Johnson's formulation of the notion of exchangeability and its relevance for Bayesian statistics.

studied engineering in Berlin and then went to Manchester as a research student, to end up in Cambridge in 1911 to attend Russell's lectures on the foundations of mathematics. That was in fact the beginning of his philosophical career, that after various vicissitudes resulted in his election as Fellow of Trinity College in 1931. With the exception of a number of trips abroad, during the second half of his life Wittgenstein lived in Cambridge. There he died in 1951, and is buried in St. Giles cemetery. Much has been written about Wittgenstein's personality and work[17], and great efforts have been made to highlight his philosophical ideas, which exercised a great influence on many thinkers, especially logical empiricists and analytical philosophers. As a matter of fact, Wittgenstein's views on probability are not very articulate, but they are nevertheless worth recalling in view of the author's notoriety.

Wittgenstein regards probability as a logical relation between propositions, analogous to that encountered in deductive logic. The latter analyses the relations between propositions on the basis of their truth values. Starting with the two values 'true' and 'false', the truth value of propositions, like '$p$ or $q$', 'not $p$', '$p$ and $q$', 'if $p$, then $q$', is determined on the basis of the value of elementary propositions, like $p$ and $q$. In the same way, Wittgenstein believes that a probabilistic relation between propositions can be established. As in the case of deductive logic, to ascertain the probability of a proposition determined on the basis of another proposition, one has to consider the 'truth-grounds' of such propositions, or the cases in which each of them can be said to be true, on the basis of a procedure like that described above. In the case of deductive logic we have that 'the truth of a proposition "$p$" follows from the truth of another proposition "$q$" if all the truth-grounds of the latter are truth-grounds of the former' (Wittgenstein 1921, English edition 1922, 1961, 2000, p. 38). The probabilistic case is the object of Proposition 5.15 of the *Tractatus*, which says that

> If $T_r$ is the number of truth-grounds of a proposition '$r$', and if $T_{rs}$ is the number of the truth-grounds of a proposition '$s$' that are at the same time truth-grounds of a proposition '$r$', then we call the ratio $T_{rs} : T_r$ the degree of *probability* that the proposition '$r$' gives to the proposition '$s$' (*ibid.*, p. 40).

There has been some discussion of whether Wittgenstein's notion of 'truth-

---

[17] See, for example, Kenny (1973) and McGuinness (1988).

ground' is related to von Kries' notion of 'Spielraum'[18]. This was suggested by Georg Henrik von Wright, with the proviso that he knew of 'no direct evidence for this' (von Wright 1982, p. 147), and denied by Brian McGuinness because 'Wittgenstein does not in fact use this notion in connexion with probability' (McGuinness 1982, p. 165), as did von Kries and Waismann. Von Wright's conjecture is endorsed by Michael Heidelberger, who makes the further guess that Wittgenstein might have come to know of von Kries' work through Boltzmann[19]. Heidelberger also considers the possibility, dismissed by von Wright, that Wittgenstein knew of Bolzano's ideas on probability, which bear a strict resemblance to his own.

An important feature of Wittgenstein's conception of probability is that it makes probability assignments depend crucially on background knowledge, including in the first place the laws of nature governing the phenomena at hand. In Wittgenstein's words: 'if I say, "The probability of my drawing a white ball is equal to the probability of my drawing a black one", this means that all the circumstances that I know of (including the laws of nature assumed as hypotheses) give no *more* probability to the occurrence of the one event than to that of the other' (Wittgenstein 1921, English edition 1922, 1961, 2000, p. 41). In the light of this, McGuinness argues that Wittgenstein does not advocate 'a Principle of Indifference based on ignorance'; quite on the contrary 'equipossibility (or some other assignment of initial probabilities) is given us in practice by the sort of laws we assume to hold in the relevant area (say in dice playing)' (McGuinness 1982, p. 165). For instance, in the case of throwing a die, we know that this phenomenon is governed by the laws of mechanics, and we know that 'they are indifferent to what result will ensue' (*ibid.*, p. 164). Wittgenstein does not relate equipossibility to independence, nor does he give a satisfactory definition of the latter, as pointed out by von Wright, who considers 'a lacuna of the *Tractatus* that it does not define or make precise this notion' (von Wright 1982, p. 140).

Wittgenstein's writings also contain a few remarks on frequency and statistical probability. In particular, the topic is addressed in the conversations recorded in 1930 by Waismann. There, after reaffirming his thesis that 'probability is an internal relation between propositions', Wittgenstein claims that 'an entirely different case is that of insurance cases. This is a matter of *a*

---

[18] On Johannes von Kries, whose approach is not discussed in the present book, the reader is addressed to Kamlah (1983). See also Kamlah (1987), where von Kries' position is compared to that of some of his contemporaries.

[19] See Heidelberger (2001).

*posteriori* probability. This has nothing to do with probability'[20] (McGuinness, ed. 1967, 1979, p. 94). In other words, statements about statistical probability are to be kept separate from proper probability statements, expressing *a priori* probabilities. The latter cannot be made to depend on frequencies, which are much too unstable to offer a safe ground for probability. As McGuinness puts it, for Wittgenstein 'a statement can be a statement of probability only if it has some independent basis in our picture of the world, some anchorage which makes it resistant from the time being to the fluctuations of frequency' (McGuinness 1982, p. 171). According to McGuinness, for Wittgenstein probability exists 'not as a feature of the world, but as a feature of our system of description. It is not a tendency in things that enables us to disregard a particular run [...] but a picture of the world to which we are independently committed' (*ibid.*, p. 170). Having pointed out this crucial feature of Wittgenstein's conception of probability, it should be added that the relationship between *a priori* and *a posteriori* probability does not seem to emerge with clarity from his writings. This may be due to the fragmentary character of his remarks on probability and frequency, some of which, particularly those contained in Waismann's recorded conversations, are likely to be comments on the work done by the same Waismann, author of a paper on probability largely inspired by Wittgenstein's ideas.

## Friedrich Waismann

Friedrich Waismann was born in Vienna in 1896 to an Austrian mother and a Russian father. After studying mathematics and physics, he approached philosophy under the guidance of Moritz Schlick, who soon became his mentor. With Schlick and Hans Hahn – his mathematics teacher – Waismann took part in the logical empiricist movement revolving around the Vienna Circle. According to the Vienna Circle historian Friedrich Stadler, Waismann 'regularly attended and organized the Vienna Circle. After its dissolution in 1936 until he emigrated, Waismann was the central figure of a discussion circle uniting former students of Moritz Schlick' (Stadler 1997, English edition 2001, p. 744). Between 1926 and 1933 he and Schlick had regular discussions with Wittgenstein, who exercised a great influence on his thought[21]. In 1937 Waismann emigrated to England. There he stayed for two years in Cambridge, then he moved to Oxford, where he was member of New College and taught

---

[20] It is worth noting that Waismann puts a question marks after this statement.
[21] These conversations are reported in McGuinness, ed. (1967).

philosophy until his death in 1959. To Waismann's disappointment, the Viennese conversations with Wittgenstein could not be resumed when he went to Cambridge because of Wittgenstein's refusal. Waismann's life was marked first by exile, by the loss of his other family in the Holocaust, and later by the tragic death first of his wife and then of his son. He is described as a deep thinker and a gentle man, who ended up conducting a secluded existence. In an obituary of Waismann, Stuart Hampshire writes that 'those who knew him, and particularly graduate and undergraduate pupils, recognized the extraordinary depth and tenacity of his inner life, and an eccentric privacy, gentleness, and detachment. He seemed to be an utterly unworldly thinker, the figure of a philosopher of romance, and he was in fact exactly what he seemed' (Hampshire 1960, p. 317).

Waismann's position regarding probability is contained in the paper 'A Logical Analysis of the Concept of Probability', read at the conference on 'The epistemology of the exact sciences' held in Prague in 1929, which was the first international meeting organised by the logical empiricist movement. The proceedings of this conference were published in 1930, in the first volume of the newly founded journal *Erkenntnis*, jointly edited by Carnap and Reichenbach. Waismann's declared intent is to proceed one step further in the direction pointed by Wittgenstein. Probability is taken to be a logical relation between statements, 'a relation which could be called the degree of "logical proximity" of two statements' (Waismann 1930, English edition 1977, p. 9). To introduce probability, Waismann starts from a parallel with deductive logic. When the scope of a proposition includes that of another, we say that the second follows from the first, or that there is a relation of entailment between them. This kind of relation can be generalized to the case in which the scope of one proposition partially overlaps with that of another[22]. Probability as a logical relation between propositions applies to this case. Waismann calls 'a relation of the kind portrayed an "internal" relation. The word "internal" is here meant to indicate that the relation is, as it were, already contained in the propositions and not just added on to them afterwards' (*ibid.*, p. 10).

Waismann proceeds to define a measure for the magnitude of a scope by fixing three conditions: (1) such a measure has to be a real, non-negative number, (2) a contradiction has measure 0, (3) given two incompatible statements, the measure of their disjunction is given by the sum of their

---

[22] According to Heidelberger, Waismann's notion of 'scope' – 'Spielraum' in the German original – can be traced back to von Kries, albeit there is no mention of this author in Waismann's paper. See Heidelberger (2001).

measures. Statements satisfying these conditions are called 'measurable'. Probability applies to measurable statements and is defined as follows: given two statements $p$ and $q$, the probability assigned to the statement $q$ by the statement $p$ is 'the magnitude of the common scope of $p$ and $q$ in proportion to the magnitude of the scope of $q$. The probability, so defined, is then, as it were, a measure of the logical proximity or deductive connection of the two propositions' (*ibid.*, p. 11).

After maintaining that 'what is probable is not the proposition, but our knowledge of the truth of a proposition', Waismann wants to safeguard his notion of probability against the charge of subjectivism. To this effect, he adds that his view 'has nothing to do with subjectivity; for what it brings out is the logical relations between propositions, and no one will want to call these subjective' (*ibid.*, p. 12). This, as we have seen, is the hallmark of logicism, as opposed to subjectivism. We shall see that a similar attitude is taken by Carnap, who retains an objective view of probability, while embracing the logical interpretation.

Waismann stresses that, in addition to a logical aspect, probability has an empirical side. In fact probability has to do with frequency, though it is '*more than a mere record of frequency*' (*ibid.*, p. 20). Unlike frequency, which is an empirically observable property of events, probability defined as a measure of the connection between propositions is logical and objective, and therefore not variable. The probability of a certain proposition can only vary if the same proposition is related to a new one, but in this case we face a new probability. The probability calculus itself is, like all of mathematics, purely logical, though it applies to frequencies. Waismann says that

> What is verified by statistical observation is always our initial assumptions, never the probability calculus itself. When a physicist deduces a consequence from his theory and confirms it by experiment, no one will want to assert that he has thereby verified the validity of logic. But in the case of the probability calculus it is thought even today that its correctness depends on experience and that it could be changed through experience. It should count as obvious that just as a physicist does not check the validity of logic with his experiments, so nothing can be settled about the theorem of addition through statistical experiences (*ibid.*, p. 14).

Not much is added on the relationship between probability and frequency, 'a clear picture' of which 'has been a long time in the making' (*ibid.*, p. 20). The conclusion reached by Waismann is that 'only by taking both elements into account can we reach a satisfactory elucidation' (*ibid.*, p. 21).

## 6.6 Rudolf Carnap's inductive logic

Two concepts of probability

The logical interpretation of probability reaches its climax with Rudolf Carnap, one of the last century's most prestigious philosophers of science. Carnap was born in Ronsdorf in 1891 and studied physics, philosophy and mathematics under Gottlob Frege, by whom he was deeply influenced. When he went to Vienna in 1925 to give a series of lectures after Schlick's invitation, he started a most fruitful collaboration with the Viennese group. Between 1926 and 1936 taught philosophy in Vienna and Prague, and became a leading figure of logical empiricism. In 1929 he wrote, together with Hans Hahn and Otto Neurath, the 'Manifesto' of the Vienna Circle[23] and in 1930 became co-editor of *Erkenntnis*, with Hans Reichenbach. In 1936 Carnap emigrated to the United States, becoming professor of philosophy first in Chicago and then in Los Angeles, where he died in 1970. Herbert Feigl describes Carnap as an 'extremely rational, calm, well-balanced' man, 'eager to absorb new information, never reluctant to change his mind even in vitally important matters'[24]. He was also a very meticulous man, a hard worker, a strong supporter of progress, peace and socialism, and a promoter of tolerance in all fields, from politics to philosophy. Carnap's ideas exercised a decisive influence on the logical empiricist movement, both at its start and during its successive phases. As a matter of fact, Carnap's thought underwent various changes and developments, and so did logical empiricism, partly under his influence.

In his intellectual autobiography, Carnap claims that his conception of probability was influenced in the first place by Wittgenstein and Waismann. Of the latter, he especially appreciated that 'his concept was not a purely logical one but combined the logical point of view of ranges with the empirical point

[23] See Carnap, Hahn and Neurath (1929).
[24] See 'Homage to Rudolf Carnap' in Hintikka, ed. (1975), p. xv, containing a collection of reminiscences of Carnap by H. Feigl, C.G. Hempel, R.C. Jeffrey, W.v.O. Quine, A. Shimony, J. Bar-Hillel, H.G. Bonhert, R.S. Cohen, C. Hartshorne, D. Kaplan, C. Morris, M. Reichenbach and W. Stegmüller.

of view of frequencies' (Carnap 1963b, p. 72). The admission of two concepts of probability is the hallmark of Carnap's approach, meant to combine a logical notion of probability with the frequentist notion developed by his Berlinese colleagues, especially his close collaborator and friend Reichenbach. Carnap started to work on probability in the Forties, although around 1936-37 he had already felt the need to develop a quantitative notion of confirmation, in connection with the problem of cognitive significance of theoretical terms that occur within the language of science. After the publication, in 1950, of *Logical Foundations of Probability*, a milestone of the literature on the philosophy of probability, Carnap devoted his attention almost exclusively to probability in the last twenty years of his life, publishing a number of important works. Although over this long period his perspective underwent significant changes, Carnap never abandoned the programme of developing an inductive logic, aimed at providing a rational reconstruction of probability within a formalized logical system. To use Richard Jeffrey's colourful expression, Carnap 'died with his logical boots on, at work on the project' (Jeffrey 1991, p. 259). In so doing, Carnap was inspired by an unwavering faith in the powers of formal logic on the one side, and of experience on the other, in compliance with the logical empiricist creed. In fact Carnap's perspective combines a belief in formal logic, as a proper tool for an explication of probability, with the conviction that one can isolate 'the experiential element in knowledge from the logical element', an attitude which according to Jeffrey is 'the distinguishing mark of empiricism' (Jeffrey 1975, p. 48).

Right at the beginning of *Logical Foundations of Probability*, Carnap sets the purpose of the work as follows:

> The theory here developed is characterized by the following *basic conceptions*: (1) all inductive reasoning, in the wide sense of nonde-ductive or nondemonstrative reasoning, is reasoning in terms of probability; (2) hence inductive logic, the theory of the principles of inductive reasoning, is the same as probability logic; (3) the concept of probability on which inductive logic is to be based is a logical relation between two statements or propositions; it is the degree of confirmation of a hypothesis (or conclusion) on the basis of some given evidence (or premises); (4) the so-called frequency concept of probability, as used in statistical investigations, is an important scientific concept in its own right, but it is not suitable as the basic concept of inductive logic; (5) all principles and theorems of inductive logic are analytic; (6) hence the validity of inductive reasoning is not dependent upon any synthetic

presuppositions like the much debated principle of the uniformity of the world (Carnap, 1950, 1962, 1967, p. v).

The admission of two notions of probability, one logical and one statistical, is clearly expressed in Carnap's papers published in the Forties, especially 'The Two Concepts of Probability' (1945a) and 'On Inductive Logic' (1945b). While recognizing the significance and utility of both concepts, the author seeks to overcome the opposition between different absolutisms, in favour of a tolerant viewpoint. In this spirit, Carnap charges both von Mises and Jeffreys with making the same mistake of regarding their own approach, frequentist in the case of von Mises and logicist in the case of Jeffreys, as the only right one. For Carnap, the logical and statistical theories of probability 'deal with two different probability concepts which are both of great importance for science. Therefore, the theories are not incompatible, but rather supplement each other' (Carnap 1945b, 1972, p. 51).

Of the two concepts of probability, the first, called *probability*$_1$, or *degree of confirmation*, is a logical concept, having to do with the degree to which a given hypothesis is confirmed by a given piece of evidence, while the second, called *probability*$_2$, refers to 'the relative frequency in the long run of one property with respect to another' (*ibidem*). A statement of probability$_2$ 'is factual and empirical, it says something about the facts of nature, and hence must be based upon empirical procedure, the observation of relevant facts' (Carnap 1945a, 1949, p. 339). On the contrary, a statement of probability$_1$ 'can be established by logical analysis alone. [...] It is independent of the contingency of facts because it does not say anything about facts' (*ibidem*). In other words, probability$_1$, or logical probability, is analytical. Calling *h* and *e* two sentences standing respectively for a certain hypothesis and a given piece of evidence, and *c* (*h, e*) the degree of confirmation of *h* based on *e*[25], one can say that, although *h* and *e* refer to facts, once they are given 'the question of confirmation requires only that we are able to understand the terms, i.e., grasp their meanings, and to discover certain relations which are based upon their meanings' (*ibid.*, p. 331). Interpreted thus, probability$_1$ becomes the object of a new logic, that Carnap calls 'inductive logic' and regards as a sort of continuation of deductive logic. 'Both branches of logic have this in common:

---

[25] Carnap claims to borrow this terminology from the Polish logician Janina Hosiasson-Lindenbaum (1899-1942), who actively worked on induction and confirmation, before being killed by the Gestapo in Vilna. For Carnap's admission see Carnap (1950), p. 23. Hosiasson's work is surveyed in Niiniluoto (1998).

solutions of their problems do not require factual knowledge but only analysis of meaning. Therefore, both parts of logic [...] belong to semantics' (*ibidem*). In this sense Carnap's work on probability can be seen as a continuation of his pioneering work in semantics, and the use of methods of formal semantics is indeed a distinctive feature of his logicism. In his writings, Carnap concentrates on probability$_1$, in the conviction that probability$_2$ has been sufficiently developed by others, like Reichenbach. The habit of calling 'Big Rudi' probability$_2$ and 'Little Rudi' probability$_1$ reflects his will to help logical probability grow as big as statistical probability[26].

Carnap points out that both concepts of probability have an objective import. As a matter of fact, in the Forties Carnap seems unable to even conceive of a subjective notion of probability. So for instance in 'The Two Concepts of Probability' he says: 'I believe that practically all authors really have an objective concept of probability in mind, and that the appearance of subjectivist conceptions is in most cases caused only by occasional unfortunate formulations' (Carnap 1945a, 1949, p. 340). Commenting on the upholders of an epistemic notion of probability, such as Laplace, Keynes and Jeffreys, he affirms that

> most and perhaps all of these authors use objectivistic rather than subjectivistic methods. [...] It appears, therefore, that the psychologism in inductive logic is, just like that in deductive logic, merely a superficial feature of certain marginal formulations, while the core of the theories remains thoroughly objectivistic (*ibid.*, p. 342).

He also praises Ramsey for holding a position similar to his own, thereby revealing a deep misunderstanding of Ramsey's ideas, because Ramsey, as argued in the next chapter, is a subjectivist, and upholds precisely that kind of psychologism that Carnap condemns.

A fundamental aspect of the distinction between probability$_1$ and probability$_2$ lies with the different sense to be attached to the expression 'unknown probability'. In fact, the value of probability$_2$ can be unknown 'in the sense that we do not possess sufficient factual information for its calculation' (*ibid.*, p. 345). Probability$_2$ represents a physical magnitude. Therefore a statement of probability$_2$ has to be established empirically, like any other statement regarding physical properties (like temperature) it can be tested in order to be confirmed or disconfirmed. Therefore probability$_2$ 'has only one

---

[26] This is reported in Salmon (1979), p. 15.

value' which is usually not known; what is known is the observed relative frequency. On the contrary, probability$_1$ cannot be said to be unknown in the same sense, though 'it may, of course, be unknown in the sense that a certain logico-mathematical procedure has not yet been accomplished, that is, in the same sense in which we say that the solution of a certain arithmetical problem is at present unknown to us' (*ibidem*). However, Carnap believes that, in principle, probability$_1$ can always be assigned a value; he therefore disagrees with Keynes' claim that there are non measurable probabilities.

Like Keynes, Carnap stresses the relative character of logical probability. Being a relation between a premiss (evidence) and a conclusion (hypothesis), probability as degree of confirmation should never be taken in isolation from the evidence supporting it. But unlike Keynes, Carnap aims at developing a general notion of logical probability, apt to be entirely reconstructed in formal terms. Moreover, Carnap thinks that there are correct values of probability$_1$, that can in principle be known and measured. He is led by this conviction to impose upon inductive logic a methodological requirement of 'total evidence' according to which 'in the application of inductive logic to a given knowledge situation, the total evidence available must be taken as basis for determining the degree of confirmation' (Carnap 1950, 1962, 1967, p. 211). The requirement of total evidence immediately raises some perplexity, since in real life the certainty of having taken into account all relevant evidence can never be reached. Carnap is aware of the problems faced by total evidence, and, after pointing out that the requirement does not belong to inductive logic proper, but rather to the methodology of induction, claims that all applications of logic to practical situations involve some degree of abstraction and approximation. In this regard he says:

> examples of the application of inductive logic must necessarily show certain fictitious features and deviate more from situations which can actually occur than in the case of deductive logic. This fact, however, does not make inductive logic a fictitious theory without relevance for science or practical life. A man who wants to calculate the areas of islands and countries begins with studying geometrical theorems illustrated by examples of simple forms like triangles, rectangles, circles, etc., although none of the countries in which he is interested has any of these forms. He knows that by beginning with simple forms he will learn a method which can be applied also to more and more complex forms approximating more and more the areas in which he is interested. Analogously, the method of inductive logic, although first

applied only to fictitious simple situations, can, if sufficiently developed, be applied to more and more complex cases which approximate more and more the situations in which we find ourselves in real life (*ibid.*, p. 213).

The ideas sketched in the Forties are fully developed in Carnap's major works of the Fifties: *Logical Foundations of Probability* (1950) and *The Continuum of Inductive Methods* (1952). In these works the interpretation of probability$_1$ and probability$_2$ remains much the same with respect to Carnap's earlier writings, though their relationship is clarified in more detail. Probability$_1$ 'has its place in inductive logic and hence in the methodology of science', probability$_2$ 'in mathematical statistics and its applications' (Carnap 1952, p. 5). Within the methodology of science, probability$_1$ plays a dual role, being used both as a method of confirmation and as a method of estimation of relative frequencies. In fact, probability$_1$ can also be seen as an estimate of probability$_2$, and this interpretation offers a way of bridging the gap between the two notions of probability. The task accomplished in *The Continuum of Inductive Methods* is precisely that of showing that there is a complete correspondence between the two meanings of probability$_1$, in the sense that there is a one-to-one correspondence between the confirmation functions and the estimate functions, and that these functions form a continuum, within the specified logical calculus.

## The logic of confirmation

Inductive logic, or the logic of confirmation functions, is developed by Carnap as the inductive analogue of deductive logic. In this section an attempt will be made to convey the basic idea underlying Carnap's inductive logic, by examining a simple example. The latter will also illustrate the probabilistic property called by Carnap 'symmetry', which corresponds to Johnson's 'permutation postulate' and to de Finetti's 'exchangeability'. Before embarking on such exposition, it is worth pointing out that, although the use of inductive logic as a logic of decision is especially stressed in his late writings, Carnap always regarded inductive logic as a guide for practical decisions. In the article 'Probability as a Guide in Life' he goes as far as claiming that inductive logic can be seen as a 'rational basis for decisions' (Carnap 1947a, pp. 147-8). This is so because inductive logic allows for making the best estimates based on the given evidence. In this sense, the logic of degrees of confirmation justifies the decisions taken on its basis, which can therefore be deemed 'rational': 'thus inductive logic serves as an instrument for the determination of rational decisions' (Carnap 1953, p. 196).

Inductive logic is developed as an axiomatic system, formalized within a first-order predicate calculus with identity, which applies to measures of confirmation defined on the semantical content of statements. Take a situation in which there are three balls, represented in the formal language of our logic by the letters $a$, $b$, and $c$, and one predicate, $P$, which stands for the property of being red[27]. The formal language also includes the negation symbol, namely '~', and the conjunction symbol, namely '·'. In this language it is possible to form various expressions, like '$Pa$', '$\sim Pb$', '$Pa \cdot Pb$'. According to the adopted interpretation, the first of these expressions means '$a$ is red', the second '$b$ is not red' and the third '$a$ and $b$ are both red'. In first order calculus, we can also formalize expressions like 'all the balls are red' and 'at least one ball is red', which are rendered as '$(x)\,Px$' and '$\exists x\,Px$'. We can also express the disjunction of statements, using the disjunction symbol '$\vee$'. Even though extremely simple, this language is good enough to convey the essentials of Carnap's endeavour. The first step to be made is to describe within our language all the possible states that can obtain, with respect to the situation under consideration. In our case there are eight possible states that can obtain, which can be described as follows:

| | |
|---|---|
| 1. $Pa \cdot Pb \cdot Pc$ | 5. $\sim Pa \cdot \sim Pb \cdot Pc$ |
| 2. $\sim Pa \cdot Pb \cdot Pc$ | 6. $\sim Pa \cdot Pb \cdot \sim Pc$ |
| 3. $Pa \cdot \sim Pb \cdot Pc$ | 7. $Pa \cdot \sim Pb \cdot \sim Pc$ |
| 4. $Pa \cdot Pb \cdot \sim Pc$ | 8. $\sim Pa \cdot \sim Pb \cdot \sim Pc$. |

These are called by Carnap *state descriptions*. The first state description tells us that $a$, $b$ and $c$ have the property $P$; the second says that $a$ does not have the property $P$, while $b$ and $c$ do; and so on. Taken together, these state descriptions convey the information on all the possible distributions that can obtain, with respect to the individuals $a$, $b$ and $c$ and the property $P$.

Any hypothesis as to our balls being red or not can be expressed in this language in terms of state descriptions. For instance, the claim $(a)$ 'the ball $a$ is red' can be expressed as the disjunction of state descriptions 1, 3, 4 and 7;

---

[27] The example is borrowed from Salmon (1966).

while the claim (*b*) 'there is at least one red ball' can be expressed as the disjunction of state descriptions 1, 2, 3, 4, 5, 6 and 7. We say that (*a*) *holds* in the specified state descriptions, and we call the ensemble of all state descriptions in which a given statement holds its *range*. In other words, the range of a given statement in our language consists of all the state descriptions that are logically compatible with it, while those state descriptions that do not fall within the range of a statement are incompatible with it. In the same way in which hypotheses can be formulated within our language, the available evidence can also be described. The degree to which a statement conveying a certain piece of evidence confirms another statement, reporting a given hypothesis, can be determined through a comparison of the ranges of the two statements. In other words, the degree to which a given piece of evidence confirms a given hypothesis corresponds within Carnap's logic to the overlapping of the ranges of the statements expressing such evidence and hypothesis. In this way, Carnap develops Waismann's idea of 'logical proximity' of two statements into a semantical concept of confirmation, defined as a relation between the meanings of the statements involved.

In order to measure degree of confirmation, one starts by assigning a weight to state descriptions. If, in a Laplacean spirit, all state descriptions were regarded as equally possible, it would be natural to assign them equal weight. In our case, each state description would weigh 1/8.

Another way of weighting state descriptions can be suggested by the consideration that, in a certain sense, descriptions 2, 3 and 4 convey similar information, namely that two out of three of our balls possess the property *P*, and one does not. The same holds for descriptions 5, 6 and 7, saying that two out of three of the balls do not have the property *P*, and one does. If one decided to stay with this kind of information, namely to take into account the number of individuals having the property *P*, irrespective of their possible permutations (or irrespective of the order in which they occur), one would end up with four possible ways of describing the situation:

I. All the balls are red. In symbols: $(x) \, Px$.

II. Two of the balls are red and one is not. In symbols: $(\exists 2x) \, Px \cdot (\exists 1x) \sim Px$.

III. Two of the balls are not red and one is. In symbols: $(\exists 1x) \, Px \cdot (\exists 2x) \sim Px$.

IV. None of the balls are red. In symbols $(x) \sim Px$.

Carnap calls these *structure descriptions*, and assigns to each of them weight 1/4. According to this assignment, each of the three state descriptions whose disjunction gives structure description II, namely 2, 3 and 4, will receive a weight of 1/12. Analogously, each of the three state descriptions whose disjunction gives structure description III, namely 5, 6 and 7, will have a weight of 1/12. This procedure corresponds to the adoption of what Carnap calls a *symmetric* way of evaluating probability.

Having defined a way of weighting descriptions, Carnap goes on to define confirmation functions in terms of such measures, which represent at the same time measures of the ranges of statements and prior probabilities. A confirmation function $c$ is defined (in the usual way of conditional probability) on the basis of the measure $m$, assigned to the hypothesis $h$ and the evidence $e$[28]:

$$c\,(h \mid e) = m\,(e \cdot h)\,/\,m\,(e).$$

Plainly, to different measures $m$ (corresponding to different weighting methods) there will correspond different confirmation functions. The functions corresponding to the weighting methods described above are called by Carnap $c^+$ and $c^*$, corresponding respectively to the measure $m^+$, which assigns equal weight to each state description, and to the measure $m^*$, which assigns equal weight to each structure description.

To see the difference between them, let us adopt the confirmation function $c^+$, and take the hypothesis '$Pa$' (namely, that the ball $a$ is red). We observe that the hypothesis in question holds in 4 state descriptions, namely 1, 3, 4 and 7, each of which weighs 1/8. The prior probability of $Pa$, calculated as that of the disjunction of these state descriptions, equals 4/8, or 1/2. Suppose next that the ball $c$ is observed to be red. The evidence $Pc$ holds in state descriptions 1, 2, 3 and 5, and the hypothesis $Pa$ holds in two of them, namely 1 and 3. Therefore the value of $c^+\,(Pa \mid Pc) = 2/4$, or 1/2. In other words, having found evidence $Pc$ does not enhance the value of confirmation of $Pa$. Moreover, if we also find that $b$ is red, and calculate $c^+\,(Pa \mid Pc \cdot Pb)$, we find once again the value 1/2, because $Pc \cdot Pb$ holds in two state descriptions, namely 1 and 4, and $Pa$ holds in one of them, namely 1. It is clear that the function $c^+$ does not allow learning from experience, since its adoption does not lead to any increase in confirmation as new evidence accumulates. This should not be surprising,

---

[28] Carnap uses the notation: $c\,(h, e)$ for confirmation functions, but here the usual notation for conditional probability is preferred.

given that the measure $m^+$, on which the function $c^+$ is based, works on the assumption of independence, underpinning the equiprobability distribution to state descriptions.

The behaviour of the symmetric function $c^*$ is quite different. In case the latter is adopted, the prior probability of the hypothesis $Pa$, which holds in structure description I, having weight 1/4, and in state descriptions 3, 4 and 7, having weight 1/12 each, is still 1/2. However, if $Pc$ is observed, the value of $c^*$ $(Pa \mid Pc)$ becomes 2/3. If in addition $Pb$ is observed, the value of confirmation of $c^*$ $(Pa \mid Pc \cdot Pb)$ increases to 3/4 (these simple calculations are left to the reader). The symmetric function $c^*$ makes learning from experience possible, by establishing a relevance connection between the observed evidence and the given hypothesis. It should be noticed that also $c^*$ is somehow based on a distribution of equal probability, but in this case it refers to structure descriptions, not to state descriptions.

For the reason described, Carnap regards the function $c^*$ as privileged, and believes that through its adoption all inferential methods can be reformulated within inductive logic. Special importance is ascribed in this context to 'predictive inference', or the inference from an observed to an unobserved sample. This is deemed 'the most important kind of inductive inference', with respect to which other kinds of inference 'may be constructed as special cases' (Carnap 1950, 1962, 1967, p. 568), including the singular predictive inference, the inference by analogy, the inverse inference, the universal inference and instance confirmation. As a matter of fact, the universal inference, or the inference from a sample to a hypothesis of universal form, raises a problem for Carnap's inductive logic, which is unable to assign to this kind of inference any value other than 0. This is due to the fact that the formulation of a universal hypothesis, like a law of nature, requires a language that includes an infinite number of individuals, and an infinite number of state descriptions. Instance confirmation, or the confirmation of a universal hypothesis through its instances, is Carnap's way to circumvent the problem. The problem of confirmation of universal laws has raised much debate: while some authors, including Jaakko Hintikka and more recently Sandy Zabell[29], have tried to modify Carnap's approach, to assign positive probability values to universal hypotheses, others have developed slightly different measures of confirmation[30].

---

[29] See Hintikka (1966) and Zabell (1997).

[30] See Kuipers (2000) and Festa (1993) for an extensive treatment of the topic.

In *The Continuum of Inductive Methods* confirmation functions are ordered in a continuum, according to Carnap's view that all such functions present both an empirical and a logical factor. The empirical factor corresponds to observed frequencies, while the logical factor 'is determined by the semantical rules' (Carnap 1945b, p. 67). A method like Reichenbach's, called by Carnap *straight rule*, takes into account the sole empirical factor, while a method like $c^+$ acts on an entirely aprioristic basis. The function $c^*$ stands somehow in the middle, for it takes into account both empirical elements (the frequency of observed instances) and aprioristic ones, by assigning equal probability to all possible structures. Carnap finds a way of expressing the twofold character of confirmation functions by means of a parameter called $\lambda$, 'in such a way that each method is uniquely and completely determined by the chosen value of $\lambda$' (Carnap 1963b, p. 74). As to the choice of the parameter value, he says that

> it seems that an observer is free to choose any of the admissible values of $\lambda$ and thereby an inductive method. If we find that the person $X$ chooses a greater value of $\lambda$ than the person $Y$, then we recognize that $X$ is more cautious than $Y$, i.e., $X$ waits for a larger class of observational data than $Y$ before he is willing to deviate in his estimate of relative frequency by a certain amount from its a priori value (*ibid.*, p. 75).

The probabilistic properties of confirmation functions are expressed by axioms, which function as restrictions to be imposed on them. A system of inductive logic is bound to include some kind of regularity axiom and an axiom of invariance, imposing some symmetry condition on confirmation functions. In addition, the system will include 'special axioms', fixing relevance conditions of some sort. Carnap worked on the definition of optimal inductive systems until the end of his life, refining his inductive logic in an attempt to make it more and more sophisticated and powerful. In his works of the Sixties inductive logic is modified in substantial ways, and at the time of his death Carnap was still working on the monograph 'A Basic System of Inductive Logic', which appeared posthumously in two parts[31].

## The turn of the Sixties

In the Sixties Carnap's notion of probability$_1$ underwent a significant change. This is clearly expressed in 'The Aim of Inductive Logic' (1962), republished in a modified and expanded version as 'Inductive Logic and Rational

---

[31] See Carnap and Jeffrey, eds. (1971) for Part I and Jeffrey, ed. (1980) for Part II.

Decisions' (1971), and in the 'Preface' to the second edition of *Logical Foundations* (1962), also republished with some modifications under the title 'Remarks on Probability' (1963a). Here, Carnap maintains that he wants to discard the interpretation of probability$_1$ as the degree of inductive support given to a hypothesis *h* by a piece of evidence *e*, retained in the first edition of *Logical Foundations*, in favour of an interpretation of it as a fair betting quotient. This move is motivated by the fact that the notion of inductive support is now seen by Carnap as ambiguous. When it is not used as a method of estimation of relative frequencies, probability$_1$ is thus interpreted as a fair betting quotient. This latter notion is taken to provide a tool for 'a rational reconstruction of the thoughts and decisions of an investigator' that 'could best be made in the framework of a probability logic' (Carnap 1963a, p. 67). In this vein, in his late writings Carnap regards inductive logic as a theory of decision. Contextually, such writings incorporate a justification of the basic principles of inductive logic in terms of coherence, typical of the subjectivistic approach of Ramsey and de Finetti. Such an appeal to coherence was suggested by Abner Shimony, who called Carnap's attention to the fact that probability$_1$, interpreted as 'the fair betting quotient for bets on *h*, given *e* as the only evidence [...] must satisfy the condition of coherence' (Shimony 1992, pp. 267, 269). Carnap gladly accepted Shimony's suggestion, because this approach offered a solution to the vexed question of the adequacy of confirmation functions, that could thereby be shown to satisfy the probability calculus[32].

Carnap's appeal to coherence has fostered the opinion that in his late writings he became a subjectivist. As a matter of fact, in the second part of 'A Basic System of Inductive Logic' Carnap labels his own position a '(modified) *subjectivist point of view*' (Carnap 1980, p. 112). In other passages of his late writings, however, we find claims like the following: 'the use of "subjective" for the concept of personal probability seems to me highly questionable' (Carnap 1971, p. 13). Between Carnap's position and that of the upholders of a genuinely subjectivist point of view, like de Finetti, there are deep divergencies, of which he seems to be well aware. In 'The Aim of Inductive Logic', after stating that his own concept of probability concerns '*rational credence*', Carnap observes that de Finetti 'says explicitly that his concept of "subjective probability" refers not to rational, but to actual beliefs', and adds: 'I find this puzzling' (Carnap 1962, 1972, p. 108). Carnap's attitude towards

---

[32] Shimony describes the occasion on which he pointed this out to Carnap as 'the one time when I made Carnap happy' (Shimony 1992, p. 269). In the same place the reader will find more details on the problem of adequacy of Carnap's confirmation functions.

inductive logic as a theory of decision is clear: the latter has to be a theory of *rational* decisions, dealing with *rational credence*. In this connection he clarifies that '*rational credence* is to be understood as the credence function of a completely rational person $X$; this is, of course, not any real person, but an imaginary, idealized person' (*ibidem*).

Carnap's distinction between *credence* and *credibility* is worth noticing. While credence reflects the beliefs of an agent at certain specified times, credibility expresses his permanent dispositions for forming and changing his beliefs in the light of information. Credibility can also be seen as the initial credence of a hypothetical human being, before experiencing empirical data. It is credibility, not credence, that, according to Carnap, provides a good basis for rational decision theory. In order to define the reasonableness of a person's credibility function, 'a sufficient number of requirements of rationality' (Carnap 1971, p. 22) have to be fixed, like coherence, regularity and symmetry. The system of inductive logic so obtained has a normative function.

Decision theory, and more specifically the approach in terms of 'beliefs, actions, possible losses, and the like' gives reasons for accepting the axioms and choosing among different credibility functions. In this way, purely logical constructs are selected on the basis of considerations that are not purely logical. As a matter of fact, in 'A Basic System of Inductive Logic, Part II', published posthumously in 1980, Carnap admits that a $\lambda$-function can be chosen on a personal basis on account of subjective and contextual elements. But in other places he warns that his theory is

> not in the field of descriptive, but of normative decision theory. Therefore in giving my reasons, I do not refer to particular empirical results concerning particular agents or particular states of nature and the like. Rather, I refer to a *conceivable* series of observations [...] to conceivable sets of possible acts, to possible states of nature, to possible outcomes of the acts, and the like. These features are characteristic for an analysis of *reasonableness* of a given function [...] in contrast with an investigation of the *successfulness* of the (initial or later) credence function of a given person in the real world. Success depends upon the particular contingent circumstances, rationality does not (Carnap 1962, 1972, p. 117).

Carnap's notion of rationality and the normative character ascribed to inductive logic, reflected by the above passage[33], sets him far apart from the subjectivism of Ramsey and de Finetti, upholders of a descriptive approach to decision theory and probability.

The stress on the rationality of inductive methods, as opposed to their successfulness, also has a bearing on Carnap's attitude to the problem of the justification of induction in his late writings. This marks a turn from his earlier perspective, decidedly sympathetic with that of Reichenbach. This is manifest in 'Probability as a Guide in Life' and 'On Inductive Logic', where he says that 'Reichenbach was the first to raise the problem of the justification of induction in a new sense and to take the first step towards a positive solution' (Carnap 1945b, 1972, p. 78). Therefore, Carnap endorses Reichenbach's idea that a viable justification of induction is based on its success. In the Fifties, once the continuum of inductive methods is constructed, the problem of justification mingles with the problem of the choice of a particular inductive method. The problem does not receive a novel solution, though Carnap seems to see the success of inductive methods as the canon for their choice. In the Sixties, and especially in the section of 'Replies and Systematic Expositions' called 'An Axiom System for Inductive Logic', and in the 1968 article 'Inductive Logic and Inductive Intuition', Carnap abandons a pragmatic approach to the problem of induction, in favour of the notion of 'inductive intuition'. To the question what reasons can be given for accepting the axioms of inductive logic, Carnap answers that 'the reasons are based upon our intuitive judgments concerning inductive validity, i.e., concerning inductive rationality of practical decisions (e.g. about bets)' (Carnap 1963c, p. 978).

The concept of inductive intuition serves to keep inductive logic entirely within the field of *a priori* knowledge. In this vein, Carnap says that the reasons for accepting the axioms that are suggested by intuition 'are a priori', they 'are independent both of universal synthetic principles about the world, e.g. the principle of the uniformity of the world, and of specific past experiences, e.g., the success of bets which were based on the proposed axioms' (*ibid.*, pp. 978-9). Whether the notion of 'inductive intuition' provides solid grounds for justifying induction is an open question. Wesley Salmon, for instance, answers this question in the negative, and remarks that Carnap's solution comes 'dangerously close to the view that induction needs no justification precisely because it is incapable of being justified' (Salmon 1967, p. 738).

---

[33] See also Carnap (1971), p. 26.

Carnap's inductive logic has shown the consequences of the application of formal logic to probability and induction. A great merit of this approach is that of clarifying the conceptual presuppositions of probabilistic inferences. Such presuppositions, in fact, become fully explicit, once they are embodied in axioms. Similarly, under this approach the difference between probabilistic inference and its methodology clearly emerges. This positive aspect of rational reconstruction performed by logical tools is counterbalanced by the shortcoming of its awkward formalism, which has not appealed to statisticians and scientists, with the drawback that Carnap's work on probability has hardly gone beyond the restricted circle of philosophers of science. Among the latter, various doubts have been raised as to the fruitfulness of inductive logic[34]. Since Carnap's methods belong to the broader family of Bayesian methods, the discussion on them has to a certain extent mingled with that on Bayesian confirmation, which represents the mainstream tendency of the literature on probabilistic confirmation. This direction, for instance, was taken by Richard Jeffrey who, after studying under Carnap in Chicago and later publishing his last manuscripts, came to the conviction that Carnap's conception of logic of confirmation as a blend of a purely logical component and a purely empirical element, should be superseded by a more eclectic approach, closer to Bruno de Finetti's subjective Bayesianism. More will be said on this in the next chapter.

## 6.7 Harold Jeffreys between logicism and subjectivism[35]

### Bayesianism

Well known for his contribution to geophysics, Harold Jeffreys (1891-1989) is also upholder of an interesting philosophy of probability and an original epistemology grounded on it. Born in Fatfield, near Durham, to a pair of schoolteachers, Jeffreys received his higher education in Newcastle and Cambridge, where he studied mathematics and science. Elected member of St. John's College in 1914, he kept this affiliation for the rest of his life, which he spent almost entirely in Cambridge, except for a few years in London, where he worked at the Meteorological Office. After being lecturer in mathematics and

---

[34] A sceptical attitude is taken among others by Suppes. See his (2002) for a discussion of this issue.
[35] This section is partly taken from Galavotti (2003).

reader in geophysics, in 1946 he became professor of astronomy and experimental philosophy at Cambridge University, and was elected to the Royal Society in 1925. Jeffreys is described as a shy, but nonetheless sociable man, with an interest in many fields, including music, choral singing, photography, psychoanalysis and travelling. He and his wife Bertha Swirles – herself educated in mathematics and physics, and lecturer at Girton College – were for many years very active in the Cambridge milieu, and interacted with the most outstanding scientists and intellectuals of the time[36].

Jeffreys is reputedly a pioneer in various fields, including celestial mechanics, seismology, and the study of the Earth. As described by Alan Cook in a long memoir,

> Harold Jeffreys stood out among the small group of pioneers who developed the physical study of the Earth from its primitive condition at the beginning of the 20th century to its state at the launch of the first *Sputnik*. [...] There have been considerable changes in the concepts and methods of geophysics from some that he established, yet the major spherically symmetrical elements of the structure of the Earth that he did so much to elucidate, are the basis for all subsequent elaboration, and generations of students learnt their geophysics from his book *The Earth* (Cook 1990, p. 303).

Jeffreys' work also left a mark in other fields, like seismology and meteorology, and, last but not least, probability. His first contribution to the subject was the outcome of a fruitful collaboration with Dorothy Wrinch, a graduate from Girton College, then lecturer at University College London, who had approached epistemological questions under the influence of Johnson and Russell. Jeffreys' interest in probability and scientific method led to the publication, in 1931, of the book *Scientific Inference*, followed in 1939 by *Theory of Probability*. In addition, he published a number of articles on the topic.

Jeffreys is an out and out Bayesian, who used to say that Bayes' theorem 'is to the theory of probability what Pythagoras' theorem is to geometry' (Jeffreys 1931, p. 7). Jeffreys was led to embrace Bayesian method by his own work as a scientist. Working in a field like geophysics, where massive data were not

---

[36] A personal and scientific portrait of Jeffreys is to be found in *Chance* IV, no. 2, (1991); see especially the personal reminiscences of Lady Jeffreys (Jeffreys' wife), the 'Recollections of a student' by Vasant S. Huzurbazar and the scientific portrait, centred on probability and statistics, by Dennis V. Lindley.

available, he became suspicious of the frequency theory, at the time the most accredited approach to probability. Moreover, in his experience as a practising scientist he was typically faced with problems of inverse probability, having to explain experimental data by means of different hypotheses, or to evaluate general hypotheses in the light of changing data. At the time Jeffreys started working on this kind of problems together with Dorothy Wrinch, Bayesian method was in disgrace among scientists and statisticians, who for the most part adhered to frequentism. Jeffreys' contribution to probability and scientific method received little appreciation by his contemporaries, and the author engaged a debate with various people, including the physicist Norman Campbell, and the statistician Ronald Fisher[37].

In three papers written with Dorothy Wrinch between 1919 and 1923[38], the lines are drawn of an inductivist programme, that Jeffreys kept firm throughout his long life, and put at the core of an epistemological viewpoint that qualifies as genuinely probabilistic. The latter rests on the idea that probability is 'the most fundamental and general guiding principle of the whole of science' (Jeffreys 1931, p. 7), and that science represents 'a branch of the subject-matter of probability' (*ibid.*, p. 219). In fact, according to Jeffreys probability should be seen as having a broader scope than science, because the process of acquiring knowledge is permeated with probability from its first steps.

At the basis of the Bayesianism of Jeffreys and Wrinch we find the assumption that all quantitative laws form an enumerable set, and their probabilities form a convergent series. This assumption – which, according to Howie, is based on an idea that occurred to Wrinch during a picnic taken with Jeffreys at Madingley Hill in Cambridge[39] – allows for the assignment of significant prior probabilities to general hypotheses. In addition, Jeffreys and Wrinch formulate a 'simplicity postulate', according to which simpler laws are assigned a greater prior probability[40]. According to its proponents, this principle corresponds to the practice of testing possible laws in order of decreasing simplicity. With this machinery, the adoption of Bayesian method is made possible.

---

[37] See Howie (2002) for a detailed reconstruction of the genesis of Jeffreys' Bayesianism, his cooperation with Dorothy Wrinch and the polemics he entertained with Fisher.
[38] See Jeffreys and Wrinch (1919), (1921) and (1923).
[39] See Howie (2002), p. 105.
[40] A discussion of the simplicity postulate is to be found in Howson (1988).

## The interpretation of probability

Jeffreys' inductivism is grounded in an epistemic view of probability, according to which probability 'expresses a relation between a proposition and a set of data' (*ibid.*, p. 9). The approach developed by Jeffreys shares the main traits of logicism, as upheld by the authors examined in the preceding sections, but in certain respects comes closer to subjectivism. Probability is for Jeffreys 'a purely epistemological notion' (Jeffreys 1955, p. 283) expressing the degree of belief entertained in the occurrence of an event on the basis of a given piece of evidence. The idea involved in this conception of probability is that of reasonable belief, corresponding to the degree of belief warranted by a certain body of evidence, which is uniquely determined. Given a set of data, Jeffreys claims, 'a proposition *q* has in relation to these data one and only one probability. If any person assigns a different probability, he is simply wrong' (Jeffreys 1931, p. 10). His conviction that a satisfactory theory has to account for 'the existence of unique reasonable degrees of belief' (Jeffreys 1939, p. 36) puts him in line with the logical interpretation of probability, which, as we have seen, involves a normative aspect and a stress on the rationality of degrees of belief. At the same time, this feature marks a crucial divergence from subjectivism, a divergence which is described by de Finetti as that between 'necessarists' who affirm and subjectivists who deny 'that there are logical grounds for picking out one single evaluation of probability as being objectively special and "correct"' (de Finetti 1970, English edition 1975, vol. 2, p. 40). As we will see in the next chapter, for de Finetti it is perfectly feasible that two people, on the basis of the same information, make different probability evaluations. In this connection, it is worth pointing out that Jeffreys was not aware of the work of de Finetti, who is never quoted in his writings.

The need to define probability in an objective way, for Jeffreys is imposed by science itself. In fact, he aims to define probability in a 'pure' way, suitable for scientific applications. It is in this spirit that Jeffreys charges the subjective interpretation of probability put forward by Frank Ramsey – with whom he consorted and shared various interests but never discussed probability[41] – of being a 'theory of expectation' rather than one of 'pure probability' (Jeffreys 1936a, p. 326), because of its link with the principle of mathematical

---

[41] According to Howie and Lindley, Jeffreys found out about Ramsey's work on probability only after Ramsey's death in 1930; see Howie (2002), p. 117 and Lindley (1991), p. 13. As mentioned in Section 7.2, Ramsey and Jeffreys shared, among other things, an interest in psychoanalysis.

expectation. In this connection, Jeffreys observes that

> the fundamental idea [of Ramsey's position] is that the values of expectations of benefit can be arranged in an order; it is legitimate to compare a small probability of a large gain with a large probability of a small gain', and continues by saying that 'the comparison is one that a business man often has to make, whether he wants or not, or whether it is legitimate or not (Jeffreys 1939, 1961, pp. 30-1).

For the scientist Jeffreys, subjective probability is a theory for business men. This is not meant as an expression of contempt, for 'we have habitually to decide on the best course of action in given circumstances, in other words to compare the expectations of the benefits that may arise from different actions; hence a theory of expectation is possibly more needed than one of pure probability' (*ibid.*, p. 326). Still, what is needed in science is a notion of 'pure probability', not the subjective notion in terms of preferences based on expectations.

In order to define probability in a 'pure' way, Jeffreys grounds it on a principle, stated by way of an axiom, which says that probabilities are comparable: 'given $p$, $q$ is either more, equally, or less probable that $r$, and no two of these alternatives can be true' (*ibid.*, p. 16). He then demonstrates that the fundamental properties of probability functions follow from this assumption. By so doing, Jeffreys qualified as one of the first to establish the rules of probability from basic presuppositions. The same kind of accomplishment was made – totally independently – by Ramsey and de Finetti. Incidentally, de Finetti made use of the same assumption in one of his articles[42].

A point of agreement between Jeffreys and subjectivists comes in connection with the idea of 'unknown probability', which is refused by both. This instead marks a disagreement with the logicism of Keynes, whose refusal 'to admit that all probabilities are expressible by numbers' (Jeffreys 1931, p. 223) is criticised by Jeffreys[43]. It is worth pointing out that, though admitting an affinity with Keynes' perspective, Jeffreys is careful to keep his own position separate from that of Keynes. In the 'Preface' to the second edition of *Theory of Probability*, he complains at having been qualified as a 'follower' of Keynes, and draws attention to the fact that Keynes' *Treatise on Probability*

---

[42] The article in question is de Finetti (1931a). More to be found in Section 7.3.

[43] See also Jeffreys (1922) where further issues of disagreement between Jeffreys and Keynes are described.

appeared after he and Dorothy Wrinch had published their first contributions to the theory of probability, drawing the lines of an epistemic approach in the same spirit as Keynes' logicism. He also points out that the resemblance between his own theory and that of Keynes is to be ascribed to the fact that both attended the lectures of William Ernest Johnson (Jeffreys 1939, 1961, p. v). Here we find more evidence of Johnson's influence on his contemporaries.

While embracing an epistemic notion of probability, Jeffreys moves a severe criticism to the frequency theory[44]. His arguments are chiefly directed against the British frequentists, particularly John Venn and Ronald A. Fisher. Jeffreys' main argument against frequentism is that 'any definition of probability that attempts to define probability in terms of infinite sets of possible observations' has to be rejected, 'for we cannot in practice make an infinite number of observations'. To instantiate such a wrong attitude, Jeffreys mentions 'the Venn limit', 'the hypothetical infinite population of Fisher' and 'the ensembles of Gibbs', which are all judged 'useless' notions (*ibid.*, p. 11). From these premises he draws the conclusion that 'no "objective" definition of probability in terms of actual or possible observations, or possible properties of the world, is admissible' (*ibidem*). As will be argued in the following pages, this relationship is reversed within Jeffreys' epistemology, where probability comes before the notions of objectivity, reality and external world (Jeffreys 1936a, p. 325).

Before we deal with Jeffreys' probabilistic epistemology in more detail, Jeffreys' views on chance and physical probability are worth recalling. Jeffreys takes *chance* to represent a 'limiting case' of everyday assignments. Chance occurs in those situations in which 'given certain parameters, the probability of an event is the same at every trial, no matter what may have happened at previous trials' (Jeffreys 1931, 1957, p. 46). For instance, chance

> will apply to the throw of a coin or a die that we previously know to be unbiased, but not if we are throwing it with the object of determining the degree of bias. It will apply to measurements when we know the true value and the law of error already. [...] It is not numerically assessable except when we know so much about the system already that we need to know no more (Jeffreys 1936a, p. 329).

Insofar as physics is concerned, Jeffreys calls attention to those fields where

---

[44] Jeffreys' criticism of frequentism is summarized in Jeffreys (1933) and (1934).

'some scientific laws may contain an element of probability that is intrinsic to the system and has nothing to do with our knowledge of it' (Jeffreys 1955, p. 284). This is the case with quantum mechanics, whose account of phenomena is irreducibly probabilistic. Unlike the probability (chance) that a fair coin falls heads, intrinsic probabilities do not belong to our description of phenomena, but to the theory itself. Jeffreys claims to be 'inclined to think that there may be such a thing as intrinsic probability. [...] Whether there is or not – he adds – it can be discussed in the language of epistemological probability' (*ibidem*). In this connection, he criticizes Carnap for admitting two notions of probability.

Jeffreys' idea that within an epistemic conception of probability there can be room for the notion of 'physical probability' is a distinctive feature of his epistemology. It may be conjectured that Jeffreys' attitude in this connection exercised some influence on Ramsey[45], who, as we will see, shared the same attitude, and outlined a notion of 'chance' and 'probability in physics' in tune with his subjective interpretation of probability as degree of belief.

## Probabilistic epistemology

Jeffreys' epistemology is rooted in a phenomenalistic view of knowledge, that he claims to derive from Ernst Mach and Karl Pearson. However, for Jeffreys 'the pure phenomenalistic attitude is not adequate for scientific needs. It requires development, and in some cases modification, before it can deal with the problems of inference' (Jeffreys 1931, p. 225). The crucial modification required with respect to Mach's perspective amounts to the introduction of probability, more precisely probabilistic inference. While 'Mach hardly considers the question of probability', Pearson obviously embodies probability in his own conception, but he 'does not go beyond Laplace's theory' (*ibidem*). With his own perspective, Jeffreys means to make definite progress with respect to both of his mentors.

On these premises, Jeffreys develops a strongly empirical and pragmatical approach to epistemology. The author's attitude can be described as constructivist, in the sense that for him such notions as 'objectivity' and 'reality' are established by inference from experience. This requires the adoption of statistical methodology, which is the core of scientific method. Concerning the notion of 'objectivity', in the 'Addenda' to the 1937 edition of

---

[45] This is argued in more detail in Galavotti (2003).

*Scientific Inference*, Jeffreys writes that

> the introduction of the word 'objective' at the outset seems [...] a fundamental confusion. The whole problem of scientific method is to find out *what* is objective (I prefer the word *real*) and all we can do is to examine possibilities in turn and see how far they coordinate with observations. Without a theory of knowledge this question has no answer; and with it the other question 'what is the external world really like?' can be left to metaphysics, where it belongs (Jeffreys 1931, 1937, p. 255).

The same idea is expressed in *Theory of Probability*, where Jeffreys says: 'I should query whether any meaning can be attached to "objective" without a previous analysis of the process of *finding out* what is objective' (Jeffreys 1939, p. 336). The process in question is inductive, and requires the use of probability. It is the very process of knowledge acquisition, which originates in our sensations and proceeds step by step to the construction of abstract notions, which lie beyond phenomena. Such notions cannot be described in terms of observables, but are nonetheless admissible and useful, because they permit 'co-ordination of a large number of sensations that cannot be achieved so compactly in any other way' (Jeffreys 1931, 1973, p. 190).

This process eventually leads to establishing 'empirical laws' or 'objective statements'. This passage is clearly inductive, for it is only after the rules of induction 'have compared it with experience and attached a high probability to it as a result of that comparison' that a general proposition can become a law. In this procedure lies 'the only scientifically useful meaning of "objectivity"' (Jeffreys 1939, p. 336). The same holds for chance attributions. A comment made by Jeffreys in the course of a discussion of Campbell's notion of chance, seems relevant in this connection. Referring to Campbell's conviction that 'the existence of chances can be shown by experiment', Jeffreys observes that 'this appears to be possible when significance tests are applicable' (Jeffreys 1936b, p. 357). In other words, chance attributions, and more generally empirical laws, are established by statistical methods.

If the notions of 'chance' and 'objectivity' acquire a definite meaning only through scientific methodology, similar considerations apply to the strictly intertwined notion of 'reality'. A useful notion of reality obtains when some scientific hypotheses receive from the data a probability which is so high, that on their basis one can draw inferences, whose probabilities are practically the same as if the hypotheses in question were certain. Hypotheses of this kind are taken as certain in the sense that all their parameters 'acquire a permanent status'. In such cases, we can assert the associations expressed by the hypotheses in question 'as an approximate rule'. According to Jeffreys, the 'scientific notion of reality' thus obtained

> is not an *a priori* one at all, but a rule of method that becomes convenient whenever we can replace an inductive inference by an approximate deductive one. The possibility of doing this in any particular case is based on experience (Jeffreys 1937, p. 69).

Like considerations apply to Jeffreys' views on causality. His proposal is to substitute the general formulation of the 'principle of causality' with 'causal analysis', as performed within statistical methodology. This starts by considering all the variations observed in a given phenomenon as random, and proceeds to detect correlations which allow for predictions and descriptions that are the more precise, the better their agreement with observations. This procedure leads to asserting laws, which are eventually accepted because 'the agreement (with observations) is too good to be accidental' (Jeffreys 1937, p. 62). The deterministic version of the principle of causality is thereby discarded, for 'exact causality in this sense remains a hypothesis; the claim that it is a result of experience is simply false. [...] It expresses a wish for exactness, which is always frustrated, and nothing more' (*ibid.*, pp. 63-4). Within scientific practice, the principle of causality is 'inverted': 'instead of saying that every event has a cause, we recognize that observations vary and regard scientific method as a procedure for analysing the variation' (Jeffreys 1931, 1957, p. 78). Granted that the fundamental problem of scientific method is that of finding a good agreement between our hypotheses and experience, Jeffreys' judgment is that in solving this problem 'causality does not help; but it turns out that the theory of probability does' (*ibidem*).

Jeffreys' position regarding scientific laws, reality and causality seems to be inspired by the same kind of pragmatism underpinning Ramsey's views on general propositions and causality, the main difference being that Ramsey's approach is more strictly analytic, whereas Jeffreys grounds his arguments on probabilistic inference and statistical methodology alone. Jeffreys' perspective is close to subjectivism also in other respects: its constructivism, the conviction that science is fallible and that there is a continuity between science and everyday life, and, last but not least, the admission that empirical information can be 'vague and half-forgotten', a fact that 'has possibly led to more trouble than has received explicit mention' (Jeffreys 1931, 1973, p. 406). As a matter of fact, the pragmatical attitude taken by Jeffreys towards epistemology is somehow at odds with his definition of probability as degree of rational belief uniquely determined by experience, and with the idea that the evaluation of probability is an objective procedure, whose application to experimental evidence obeys rules having the status of logical principles. As suggested by the above mentioned criticism of Ramsey, Jeffreys was presumably led to this interpretation of probability by the desire to keep the definition of probability free from whatever reference to the notion of mathematical expectation.

# 7

## *The subjective interpretation*

### 7.1 The beginnings

William Donkin

William Fishburn Donkin (1814-1869), was born at Bishop Burton, Yorkshire, and educated at St. Peter's school, York. In 1832 he entered St. Edmund Hall, Oxford, and ten years later became Savilian Professor of astronomy at Oxford, a post he retained throughout his life. In his lifetime, Donkin gained a good reputation, was elected a member of the Royal Society and contributed a number of articles to leading scientific journals. Since his youth he had had a gift for mathematics, as well as for classics and music. He was reputed to be an expert on Greek music, and wrote some essays on the subject.

Donkin entrusted his concept of probability to the note 'On Certain Questions Relating to the Theory of Probabilities', published in 1851. On its first page, he writes:

It will, I suppose, be generally admitted, and has often been more or less explicitly stated, that the subject-matter of calculation in the mathematical theory of probabilities is quantity of belief. Every problem with which the theory is concerned is of the following kind. A certain number of hypotheses are presented to the mind, along with a certain quantity of information relating to them: In what way ought belief to be distributed among them? (Donkin 1851, p. 353).

Deeply convinced of the epistemic meaning of probability, Donkin adds that 'the "probability" which is estimated numerically means merely "quantity of belief", and is nothing inherent in the hypothesis to which it refers' (*ibid*, p. 355). These claims must have impressed Frank Ramsey, who recorded them in his notes[1].

Donkin's position is actually quite similar to that of De Morgan, especially when he maintains that probability is '*relative* to a particular state of knowledge or ignorance; but [...] it is *absolute* in the sense of not being relative to any individual mind; since, the same information being presupposed, all minds *ought* to distribute their belief in the same way' (*ibidem*). If claims of this kind bring Donkin closer to the logicists than to the subjectivists, the appearance of his name in the present chapter on subjectivism does not seem out of place, in view of the fact that he addresses the issue of belief conditioning in a very modern way, which anticipates the work of Richard Jeffrey a hundred years later. In this connection, Donkin formulates a principle, imposing a symmetry restriction on updating belief, as new information is obtained. In a nutshell, it states that changing opinion on the probabilities assigned to a set of hypotheses, after new information has been acquired, has to preserve the proportionality among the probabilities assigned to the considered options. Under this condition, the new and old opinions are comparable. The principle is introduced by Donkin as follows:

> Theorem. – If there be any number of mutually exclusive hypotheses, $h_1$, $h_2$, $h_3$ ..., of which the probabilities relative to a particular state of information are $p_1$, $p_2$, $p_3$ ..., and if new information be gained which changes the probabilities of some of them, suppose of $h_{m+1}$ and all that follow, without having otherwise any reference to the rest, then the probabilities of these latter have the same ratios to one another, after the new information, that they had before; that is,
>
> $$p'_1 : p'_2 : p'_3 : ... : p'_m = p_1 : p_2 : p_3 : ... : p_m,$$
>
> where the accented letters denote the values after the new information has been acquired (*ibid.*, p. 356).

As we shall see in Section 7.4 of this chapter, the method of conditioning usually called 'Jeffrey conditionalization', reflects precisely the intuition Donkin committed to the above mentioned principle[2].

---

[1] See document 003-13-01 of the Ramsey Collection, held at the Hillman Library, University of Pittsburgh.
[2] See Jeffrey (1965), and the collection of essays in Jeffrey (1992).

# Émile Borel

Born in Saint-Affrique to a protestant village pastor, Émile Borel (1871-1956) was early known to be a prodigy, and moved to study in Paris. In 1889 he entered the École Normale, where he spent many years as a researcher, and in 1909 moved to the Sorbonne, where he stayed until his retirement in 1940. Borel had an extremely active life, not only as a mathematician, but also as a politician. Deeply interested in pedagogical and social problems, he professed socialist ideas. Among other things, he was mayor of his home town Saint-Affrique, was for twelve years member of the Chamber of Deputies, and in 1925 served as minister of the Navy. His activity as a scientist and public figure earned him ample recognition and many honours from the most prestigious institutions. A successor to Poincaré's chair of probability theory and mathematical physics at the Sorbonne University in Paris, Borel made substantial contributions to twentieth century mathematics, especially to analysis and probability. He left a wide production, from technical to popular writings, on subjects ranging from the theory of functions and measure theory to game theory – a subject on which he published a series of not very well known papers between 1921 and 1927.

In a review of Keynes' *Treatise* originally published in 1924, and later reprinted in the last volume of the series of monographs edited by Borel under the title *Traité du calcul des probabilités et ses applications* (1939)[3], Borel raises various objections to Keynes' theory of probability, and outlines his own viewpoint, earlier expounded in a series of more technical works. Borel's main discontent with Keynes' work lies with the conviction that it overlooks the applications of probability to science, to focus only on the probability of judgments:

> For Mr Keynes, who is so attached to realities, to what is a 'matter of fact', are not properties of gases, or of things that dissolve, or of emulsions, or of radioactive substances, or general properties which the use of the calculus of probabilities permits us to make precise predictions about, also realities? Or are the successful applications of the calculus of probabilities enough to deprive them of reality in the eyes of Mr Keynes? Rather, what seems to be the essential question for him is the question of knowing whether, when I take a lottery ticket,

---

[3] The *Traité* includes 18 issues, collected in 4 volumes. The review of Keynes' *Treatise* appears in the last issue, under the title 'Valeur pratique et philosophie des probabilités'.

one must speak of the probability that I win, or of the probability of the judgment which I enunciate in declaring that I will win (Borel 1924, English edition 1964, pp. 47-8).

Borel takes this to be a distinctive feature of English literature, as opposed to continental, which he regards as more aware of the developments of science, particularly physics. When making such claims, Borel is likely to have in mind above all Henri Poincaré, whose ideas exercised a certain influence on him[4].

Borel claims to agree with Keynes in taking probability in its epistemic sense, to mean the probability of individual judgment, based on a given piece of evidence. But he stresses that probability acquires a different meaning depending on the context in which it occurs. In particular, he draws attention to those cases

where one refers to the probability which is common to the judgments of all the best informed persons, that is to say, the persons possessing all the information that it is humanly possible to possess at the time of the judgments. [...] The probability that an atom of radium will explode tomorrow is, for the physicist, a constant of the same kind as the density of copper or the atomic weight of gold. These constants are always at the mercy of the progress of physical-chemical theory; they are, however, constants in the present state of science (*ibid.*, p. 50).

In other words, probability has a different value in science and everyday life, and more generally in situations characterized by a different state of information. Typically, probability has a somewhat more objective meaning within science, where its assessment is grounded on a strong body of information, shared by the scientific community. Referring to the probabilities mentioned in the passage quoted above, Borel claims that

One could, in order to abridge our language, while at the same time not attaching an absolute sense to this expression, call these probabilities objective probabilities. On the other hand, subjective probabilities, those which particularly interest Mr Keynes, can effectively have different values for different individuals (*ibidem*).

The last assertion leads us straight to Borel's subjectivism, where he explicitly

---

[4] See von Plato (1994), p. 36, where Borel is described as a successor of Poincaré 'in an intellectual sense'. The book by von Plato contains a detailed exposition of Borel's ideas on probability. See also Knobloch (1987).

admits that two people, in the presence of the same information, can come up with different probability evaluations. This is most common in everyday applications of probability, like horse races, or weather forecasts. In all such cases, probability judgments are of necessity relative to 'a certain body of knowledge', which is not the kind of information shared by everyone, like scientific theories at a certain time. Now, Borel makes it clear that when talking of a probability of this kind, the 'body of knowledge' in question should be thought of as 'necessarily included in a determinate human mind, but not such that the same abstract knowledge constitutes the same body of knowledge in two distinct human minds' (*ibid.*, p. 51). Since this features a distinctive trait of subjectivism, it seems worthwhile recollecting Borel's exemplification:

> If we plan a ride in an open car or boat we instinctively ask ourselves what the probability is that rain or tempest will begin before the end of the trip. Though two persons *A* and *B* both know the state of the sky, the wind, the sea, even the variations of the barometer, it can still happen that *A* is mistaken less often than *B*, that is to say, that the predictions of *A* are more probable than the predictions of *B*. This example is quite interesting, for it can well happen that the person *A* whose prediction is surer is a peasant or a sailor, quite uneducated, but knowing well the region where he lives. Under these conditions the probability of the judgment that *A* makes about the weather will be found to be perhaps a little modified if the knowledge of *A* becomes more extended, if he learns, for example, to read daily the graphs of barometric pressure, but it will not be the same probability even in the case where his judgment remains the same, because this judgment will be related to a different body of knowledge (*ibidem*).

Probability evaluations made at different times, based on different information, ought not be taken as refinements of previous judgments, but as totally new ones. While sharing this conviction with Keynes, Borel disagrees with the latter on the claim that there are probabilities which cannot be evaluated numerically.

It is in this connection that Borel appeals to the method of betting, which 'permits us in the majority of cases a numerical evaluation of probabilities' (*ibid.*, p. 57). This method – which, as we have seen, dates back to the origin of the numerical notion of probability in the seventeenth century – is regarded by Borel as having exactly the same characteristics as the evaluation of prices by the method of exchange. If one desires to know the price of a ton of coal, it suffices to offer successively greater and greater sums to the person who possesses the coal; at a certain sum he will decide to sell it. Inversely if the possessor of the coal offers his coal, he will find it sold if he lowers his demands sufficiently (*ibidem*).

At the end of a discussion of the method of bets, where he takes into account some of the traditional objections against it, Borel concludes that this method seems good enough, in the light of ordinary experience.

Borel's conception of epistemic probability has a strong affinity with the subjective interpretation developed by Ramsey and de Finetti. In a brief note on Borel's work, de Finetti praises Borel for pleading the idea that probability must be referred to the single case, and for holding that this kind of probability is always measurable sufficiently well by means of the betting method. At the same time, de Finetti expresses strong disagreement with the eclectic attitude taken by Borel, more particularly with his admission of an objective meaning of probability in addition to the subjective – a position that, as we shall see, de Finetti always rejected[5].

## 7.2 Frank Plumpton Ramsey and the notion of coherence

### Degrees of belief and consistency

Born in Cambridge in 1903 to a family of a distinguished academic background – his father, a mathematician, was President of Magdalene College – Frank Plumpton Ramsey died of jaundice at Guy's Hospital in London in 1930 – at the age of 26 – leaving an extraordinary intellectual heritage, still to be explored in all details. After graduating in mathematics in 1923 from Trinity College, he became a Fellow of King's in 1924, and subsequently lecturer of mathematics at Cambridge. Uncommonly precocious and gifted, during his short life Ramsey made outstanding contributions to a number of different fields, including mathematics, logic, philosophy, probability, and economics. Keynes refers to Ramsey's as 'one of the brightest minds of our generation' (Keynes 1930, 1972, p. 336), and portrays 'his bulky

---

[5] De Finetti's commentary on Borel is to be found in de Finetti (1939).

Johnsonian frame, his spontaneous gurgling laugh, the simplicity of his feelings and reactions, half-alarming sometimes and occasionally almost cruel in their directness and literalness, his honesty of mind and heart, his modesty, and the amazing, easy efficiency of the intellectual machine which ground away behind his wide temples and broad, smiling face' (*ibidem*).

As a member of the Cambridge intellectual community, Ramsey interacted with his contemporaries, often influencing their ideas. Among them Keynes, Moore, Russell and Wittgenstein – whose *Tractatus* he translated into English. A regular attender at the meetings of the 'Moral Sciences Club' and the discussion group called the 'Apostles'[6], Ramsey cultivated many interests, including psychoanalysis[7], music, literature, politics and social welfare – an interest that was transmitted to him by his mother[8].

Ramsey's work prepares the ground for modern subjectivism, sometimes also called 'personalism'. His most important contribution to the subject is the well known paper 'Truth and Probability', read at the 'Moral Sciences Club' in Cambridge in 1926, then published posthumously in 1931 in the collection *The Foundations of Mathematics and Other Logical Essays*, edited by Richard Bevan Braithwaite shortly after Ramsey's death. Other sources are to be found in the same book, as well as in the other collection, edited by Hugh Mellor, *Philosophical Papers* (largely overlapping Braithwaite's), and in the volumes *Notes on Philosophy, Probability and Mathematics*, edited by Maria Carla Galavotti, and *On Truth*, edited by Nicholas Rescher and Ulrich Majer.

For Ramsey, probability is a degree of belief, and probability theory is a

---

[6] Some remarks on this group of intellectuals are to be found in Levy (1979) and Harrod (1951). The collection *Notes on Philosophy, Probability and Mathematics* (see Ramsey 1991a) includes seven papers that Ramsey read to the Apostles in the years 1921-25. Also the 'Epilogue' closing the collection *The Foundations of Mathematics* (1931), reprinted in *Philosophical Papers*, (1990a) was originally read in 1925 to the Apostles.

[7] Howie reports that Ramsey took part with Harold Jeffreys in a group discussing psychoanalysis (Howie 2002, p. 117). The activity of this group is described in some detail in Cameron and Forrester (2000). In 1924 Ramsey was psychoanalysed in Vienna by Theodor Reik, a disciple of Freud (see the Ramsey Archive at King's College, Cambridge).

[8] For some information on Ramsey's life, see the last chapter of Sahlin (1990). See also 'Better than the Stars', a radio portrait of Frank Ramsey written and presented by Hugh Mellor, with Alfred J. Ayer, Richard B. Braithwaite, Richard C. Jeffrey, Michael Ramsey (Archbishop of Canterbury and Frank's brother), Lettice Ramsey (Frank's widow), Ivor A. Richards, originally recorded in 1978, and later published in Mellor, ed. (1995). More to be found in the Ramsey Archive of King's College, Cambridge.

logic of partial belief. The notion of *degree of belief* is taken as a primitive notion, which admittedly 'has no precise meaning unless we specify more exactly how it is to be measured' (Ramsey 1990a, p. 63). In other words, to give probability a univocal meaning, an operative definition has to be produced, specifying a way of measuring it. A first option to achieve this purpose is the method of bets, endowed with a long-standing tradition: 'the old established way of measuring a person's belief is to propose a bet, and see what are the lowest odds which he will accept' (*ibid.*, p. 68). Even though 'fundamentally sound', this method 'suffers from being insufficiently general, and from being necessarily inexact' (*ibidem*). Ramsey has in mind the problem of diminishing marginal utility of money, arising from the fact that the value of winning a certain sum, say 500 dollars, is obviously much higher for an academic living on his salary, than for a tycoon like Bill Gates, and it is even higher for someone living on unemployment benefit. He also mentions the arbitrariness of the method, due to personal 'eagerness or reluctance to bet', and the fact that 'the proposal of a bet may inevitably alter' a person's 'state of opinion' (*ibidem*).

In order to avoid such difficulties, Ramsey adopts a different method, based on the notion of 'preference'. At its roots we find the conviction that 'we act in the way we think most likely to realize the objects of our desires, so that a person's actions are completely determined by his desires and opinions' (*ibid.*, p. 69). Though not fully adequate to represent all aspects of human conduct, this theory can be regarded as a

> useful approximation to the truth particularly in the case of our self-conscious or professional life, and it is presupposed in a great deal of our thought. It is a simple theory and one which many psychologists would obviously like to preserve by introducing unconscious opinions in order to bring it more into harmony with the facts (*ibidem*).

Once this theory is assumed as a general framework, apt to give a psychological foundation to the notion of partial belief, Ramsey adds that 'we seek things which we want, which may be our own or other people's pleasure, or anything else whatever, and our actions are such as we think most likely to realize these goods' (*ibidem*). He then clarifies that he does not speak of 'good' and 'bad' in an ethical sense, 'but simply as denoting that to which a given person feels desire and aversion' (*ibid.*, p. 70).

Ramsey makes the assumption that such goods are measurable, as well as additive, and furthermore assumes that an agent 'will always choose the course of action which will lead in his opinion to the greatest sum of good' (*ibidem*). In order to account for the fact that people act under uncertainty, because they hardly ever entertain a belief with certainty, Ramsey appeals to the principle of mathematical expectation, taken 'as a law of psychology' (*ibidem*). Given a person who is prepared to act in order to achieve some good,

> if $p$ is a proposition about which he is doubtful, any goods or bads for whose realization p is in his view a necessary and sufficient condition enter into his calculation multiplied by the same fraction, which is called the 'degree of his belief in $p$'. We thus define degree of belief in a way which presupposes the use of mathematical expectation (*ibidem*).

Ramsey also suggests an alternative definition of degree of belief, equivalent to the above:

> Suppose [the] degree of belief [of a certain person] in $p$ is $m/n$; then his action is such as he would choose it to be if he had to repeat it exactly $n$ times, in $m$ of which $p$ was true, and in the others false (*ibidem*).

The two accounts point out two different, albeit strictly intertwined, aspects of the same concept, and can be taken to be equivalent. A typical situation involving a choice of action that depends on belief would be the following:

> I am at a cross-roads and do not know the way; but I rather think one of the two ways is right. I propose therefore to go that way but keep my eyes open for someone to ask; if now I see someone half a mile away over the fields, whether I turn aside to ask him will depend on the relative inconvenience of going out of my way to cross the fields or of continuing on the wrong road if it is the wrong road. But it will also depend on how confident I am that I am right; and clearly the more confident I am of this the less distance I should be willing to go from the road to check my opinion. I propose therefore to use the distance I would be prepared to go to ask, as a measure of the confidence of my opinion (*ibid.*, pp. 70-1).

Denoting by $f(x)$ the disadvantage of walking $x$ metres, by $r$ the advantage of reaching the right destination, and by $w$ the disadvantage of arriving at a wrong destination, if I were ready to go a distance $d$ to ask, the degree of belief that I am on the right road is $p = 1 - (f(d)/(r-w))$. To choose an action of this kind can be considered advantageous if, were I to act $n$ times in the same way, $np$ times out of these $n$ I was on the right road (otherwise I was on the wrong one).

In fact, the total good of not asking each time is $npr + n(1-p)w = nw + np(r-w)$; while the total good of asking each time (in which case I would never go wrong) is $nr - nf(x)$. The total good of asking is greater than the total good of not asking, provided that $f(x) < (r-w)(1-p)$. Ramsey concludes that the distance $d$ is connected with my degree of belief, $p$, by the relation $f(d) = (r-w)(1-p)$, which amounts to $p = 1 - (f(d)/(r-w))$, as stated above. He then observes that

> It is easy to see that this way of measuring beliefs gives results agreeing with ordinary ideas. [...] Further, it allows validity to betting as means of measuring beliefs. By proposing to bet on $p$ we give the subject a possible course of action from which so much extra good will result to him if $p$ is true and so much extra bad if $p$ is false (*ibid.*, p. 72).

However, given the difficulties connected with the betting scheme – such as the already mentioned problem of the diminishing marginal utility of money, and the undesirable consequences of certain psychological attitudes, like aversion or propensity to risk – Ramsey turns to a more general notion of 'preference'.

Degree of belief is then operationally defined with direct reference to personal preferences, determined on the basis of the expectation of an individual of obtaining certain goods, not necessarily of a monetary kind. The value of such goods is always relative, as they are defined with reference to a set of alternatives. The definition of degree of belief is committed to a set of axioms, indicating how to represent its values by means of real values. Degrees of belief obeying such axioms are called *consistent*. Ramsey goes on to spell out the laws of probability in terms of degrees of belief, and argues that consistent sets of degrees of belief satisfy the laws of probability. Additivity is assumed in a finite sense, since the set of alternatives taken into account is finite. On this point, Ramsey comments that the human mind is only capable of contemplating a finite number of alternatives open to action, and even when a question is conceived, allowing for an infinite number of answers, these have to be lumped 'into a finite number of groups' (*ibid.*, p. 79). The laws of probability are then shown to be

> necessarily true of any consistent set of degrees of belief. Any definite set of degrees of belief which broke them would be inconsistent in the sense that it violated the laws of preference between options. [...] If anyone's mental condition violated these laws, his choice would depend on the precise form in which the options were offered him, which

would be absurd. He could have a book made against him by a cunning better and would then stand to lose in any event.

We find, therefore, that a precise account of the nature of partial belief reveals that the laws of probability are laws of consistency. [...]

Having any definite degree of belief implies a certain measure of consistency, namely willingness to bet on a given proposition at the same odds for any stake, the stakes being measured in terms of ultimate values. Having degrees of belief obeying the laws of probability implies a further measure of consistency, namely such a consistency between the odds acceptable on different propositions as shall prevent a book being made against you (*ibid.*, p. 78).

The crucial link between probability and degree of belief, provided by consistency (or coherence – to use a more widespread term) is the cornerstone of subjective probability. In fact, consistency guarantees the applicability of the notion of degree of belief, which can therefore qualify as an admissible interpretation of probability. By showing that from the assumption of coherence one can derive the laws of probability Ramsey paves the way to a fully-fledged subjectivism. Remarkably, within this perspective the laws of probability 'do not depend for their meaning on any degree of belief in a proposition being uniquely determined as the rational one; they merely distinguish those sets of beliefs which obey them as consistent ones' (*ibidem*). Coherence is the only condition that degrees of belief should obey: insofar as a set of degrees of belief is coherent, there is no further demand of rationality to be met. This marks a sharp difference between subjectivism and the logical interpretation of probability of Carnap and Jeffreys.

An important consequence of the adoption of a notion of probability in terms of coherent degrees of belief is that Ramsey does not need to ground his own theory on the principle of indifference. In this connection, he observes that 'the Principle of Indifference can now be altogether dispensed with. [...] To be able to turn the Principle of Indifference out of formal logic is a great advantage; for it is fairly clearly impossible to lay down purely logical conditions for its validity, as is attempted by Mr Keynes' (*ibid.*, p. 85). This is a decisive step in the moulding of modern subjectivism. Almost in the same years in which Ramsey was developing his ideas on probability, a further step was made by Bruno de Finetti, who supplied the 'static' definition of subjective probability in terms of coherent degrees of belief with a 'dynamic' dimension, obtained by joining subjective probability with exchangeability, within the framework of the Bayesian method[9]. Though this crucial step was actually made by de Finetti, there is evidence that Ramsey also knew the property of exchangeability, of which he must have heard from Johnson's lectures. In fact, he left a note called 'Rule of Succession', where he made use of the notion of exchangeability, named by him 'equiprobability of all permutations'[10]. What Ramsey did not see, and was instead seen by de Finetti, is the usefulness of applying exchangeability to the inductive procedure, modelled upon Bayes' rule. Remarkably enough, in another note called 'Weight or the Value of Knowledge'[11], Ramsey was able to prove that collecting evidence pays in expectation, provided that acquiring the new information is free, and shows how much the increase in weight is. This shows he had a dynamical view at least of this important process[12].

## Ramsey, Keynes and Wittgenstein

Ramsey puts forward his theory of probability in open contrast with that of Keynes. As mentioned in Chapter 6, Ramsey criticizes the hypothesis of limited variety, which he regards as ill-founded, mainly because it is based on an implicit equiprobability assumption. In addition, he does not share Keynes' conviction that 'a probability may [...] be unknown to us through lack of skill in arguing from given evidence' (Ramsey 1922, 1989, p. 220). Indeed, the very idea of unknown probability is alien to the subjectivist point of view. As we shall see, this is also emphasized by de Finetti.

Ramsey's criticism reaches the core of Keynes' perspective, with a refusal to attach a definite meaning to the logical relations, on which the latter rests. In his note 'Criticism of Keynes' Ramsey writes:

There are no such things as these relations.
a) Do we really perceive them? Least of all in the simplest cases when

---

[9] This terminology is borrowed from Zabell (1991), containing useful remarks on Ramsey's contribution to subjectivism. For a comparison between Ramsey and de Finetti on subjective probability, see Galavotti (1991).

[10] See Ramsey (1991a), pp. 279-81. What Ramsey does in this note is described in some detail in Di Maio (1994).

[11] See Ramsey (1990b); also in (1991a), pp. 285-7.

[12] Nils-Eric Sahlin's 'Preamble' to Ramsey (1990b) contains useful comments on Ramsey's note, and some bibliography on the result it contains, which anticipates subsequent work by L.J. Savage, I.J. Good, and others. See also Skyrms (1990), especially pp. 93-6, for a discussion of Ramsey's result.

they should be clearest; can we really know them so little and yet be so certain of the laws which they testify? [...]
c) They would stand in such a strange correspondence with degrees of belief' (Ramsey 1991a, pp. 273-4).

Ramsey's way of looking at the relationship between logic and probability is utterly different from that of Keynes. He distinguishes between a 'lesser logic, which is the logic of consistency, or formal logic', and a 'larger logic, which is the logic of discovery, or inductive logic' (Ramsey 1990a, p. 82). While the lesser logic, which is the logic of tautologies in Wittgenstein's sense, can be 'interpreted as an objective science consisting of objectively necessary propositions' (*ibid.*, p. 83), the larger logic, which includes probability, does not share this feature, because 'when we extend formal logic to include partial beliefs this direct objective interpretation is lost' (*ibidem*). For Ramsey, the larger logic can only be endowed with a psychological foundation. This move towards psychology turns out to be in agreement with Wittgenstein.

In a paper read to the Apostles in 1922, called 'Induction: Keynes and Wittgenstein', Ramsey addresses the problem of induction, and endorses Wittgenstein's psychologism against Keynes logicism. At the beginning of the paper, he mentions propositions 6.363 and 6.3631 of Wittgenstein's *Tractatus*, where it is said that the process of induction 'has no logical foundation but only a psychological one' (Ramsey 1991a, p. 296). After praising Wittgenstein for his appeal to psychology in order to find a justification of the inductive procedure, Ramsey discusses at length Keynes' approach, expressing serious doubts on his attempt at grounding induction on logical relations and hypotheses. At the end of the paper, after recalling Hume's celebrated argument, Ramsey puts forward by way of a conjecture, of which he claims to be too tired 'to see clearly if it is sensible or absurd', the idea that induction could be justified by saying that

> a type of inference is reasonable or unreasonable according to the relative frequencies with which it leads to truth and falsehood. Induction is reasonable because it produces predictions which are generally verified, not because of any logical relation between its premises and conclusions. On this view we should establish by induction that induction was reasonable, and induction being reasonable this would be a reasonable argument (*ibid.*, p. 301).

Here Ramsey seems to envisage a pragmatic justification of the inductive procedure. The same attitude is resumed at the end of 'Truth and Probability', where he describes his own position as 'a kind of pragmatism', and holds that

> we judge mental habits by whether they work, i.e. whether the opinions they lead to are for the most part true, or more often true than those which alternative habits would lead to.
>
> Induction is such a useful habit, and so to adopt it is reasonable. All that philosophy can do it to analyse it, determine the degree of its utility, and find on what characteristics of nature it depends. An indispensable means for investigating these problems is induction itself, without which we should be helpless. In this circle lies nothing vicious. It is only through memory that we can determine the degree of accuracy of memory; for if we make experiments to determine this effect, they will be useless unless we remember them (Ramsey 1990a, p. 93-4).

Pragmatism is a major feature of Ramsey's perspective on probability and on philosophy at large – as testified by the references in his works to William James and Charles Sanders Peirce. Among Peirce's ideas to which Ramsey was sympathetic, are the view of 'logic as self-control'[13], and the notion of scientific truth as general consent, to which we will come back in what follows.

## Belief, frequency and 'probability in physics'[14]

Well aware that, by the time he wrote 'Truth and Probability', the frequency theory was the most popular view of probability among statisticians and physicists, Ramsey felt the need to address the problem of the relationship between the frequency interpretation of probability and probability as degree of belief. He writes that

> It is natural [...] that we should expect some intimate connection between these two interpretations, some explanation of the possibility of applying the same mathematical calculus to two such different sets of phenomena (*ibid.*, p. 83).

Such a connection is identified with the fact that

> the very idea of partial belief involves reference to a hypothetical or ideal frequency [...] belief of degree $m/n$ is the sort of belief which leads to the action which would be best if repeated $n$ times in $m$ of which the proposition is true' (*ibid.*, p. 84).

---

[13] See 'Reasonable Degree of Belief', in Ramsey (1990a), p. 99.
[14] This section is partly taken from Galavotti (1999).

This passage – echoing the previously mentioned 'conjecture' from 'Induction: Keynes and Wittgenstein' – reaffirms Ramsey's pragmatical tendency to refer belief to action, and to justify inductive behaviour with reference to successful conduct. The argument is pushed even further, when Ramsey says that

> It is this connection between partial belief and frequency which enables us to use the calculus of frequencies as a calculus of consistent partial belief. And in a sense we may say that the two interpretations are the objective and subjective aspects of the same inner meaning, just as formal logic can be interpreted objectively as a body of tautology and subjectively as the laws of consistent thought (*ibidem*).

However, in other passages the connection between these two 'aspects' is not quite so strict:

> experienced frequencies often lead to corresponding partial beliefs, and partial beliefs lead to the expectation of corresponding frequencies in accordance with Bernoulli's Theorem. But neither of these is exactly the connection we want; a partial belief cannot in general be connected uniquely with any actual frequency (*ibidem*).

To sum up, the relationship between degree of belief and frequency is an open problem within Ramsey's perspective. As we will see, this problem finds an answer in de Finetti's Representation theorem, but this was not available to Ramsey[15].

Claims like the one quoted above, to the effect that the frequency interpretation and that in terms of degrees of belief 'are the two objective and subjective aspects of the same inner meaning', might be taken to suggest that Ramsey, like Carnap, admitted of two notions of probability: one epistemic (the subjective view) and one empirical (the frequency view). All the more since at the very beginning of 'Truth and Probability', Ramsey claims that although the paper deals with the logic of partial belief, 'there is no intention of implying that this is the only or even the most important aspect of the subject', adding that 'probability is of fundamental importance not only in logic but also in statistical and physical science, and we cannot be sure beforehand that the most useful interpretation of it in logic will be appropriate in physics also' (*ibid.*, p. 53). As a matter of fact, Ramsey took very seriously the problem of what kind of probability is employed in science. We know from Braithwaite's

---

[15] For some remarks on this point, see Galavotti (1991) and (1995).

'Introduction' to *The Foundations of Mathematics* that Ramsey had planned to write a final section of 'Truth and Probability', dealing with probability in science. We also know from Ramsey's unpublished notes[16] that in the last years of his short life he had been working on a book bearing the title 'On Truth and Probability', of which he left a number of tables of contents. Of the projected book he only wrote the first part, dealing with the notion of truth, which has been published under the title *On Truth*[17]. One can speculate that he meant to include in the second part of the book the content of the paper 'Truth and Probability', plus some additional material on probability in science. The notes published in *The Foundations of Mathematics* under the heading 'Further Considerations'[18], and a few more published in the volume *Notes on Philosophy, Probability and Mathematics*, contain evidence that in the years 1928-29 Ramsey was actively thinking about such problems as theories, laws, causality, chance, all of which he saw as strictly connected. A careful analysis of such writings shows that – contrary to the widespread opinion that he was a dualist with regard to probability – in the last years of his life Ramsey was developing a view of 'chance' and 'probability in physics' fully compatible with his subjective interpretation of probability as degree of belief.

Ramsey moves from the conviction that there are cases in which the evaluation of probability is guided by scientific theories. In this vein, he objects to Keynes that 'anyone who tries to decide by Keynes' methods what are the proper alternatives to regard as equally probable in molecular mechanics, e.g. in Gibbs' phase-space, will soon be convinced that it is a matter of physics rather than pure logic' (Ramsey 1990a, p. 85). This leads Ramsey to maintain that the definition of chance involves reference to scientific theories. However, chance cannot be defined simply in terms of laws or frequencies – though the specification of chances involves reference to laws, in a way that will soon be clarified. 'We sometimes really assume a *theory* of the world with laws and chances and mean not the proportion of actual cases but what is chance on our theory' writes Ramsey in 'Reasonable Degree of Belief' (*ibid.*, p. 97). The same point is emphasized in the note 'Chance', also written in 1928, where Ramsey criticizes the frequency-based views of chance put forward by other authors, like Norman Campbell. The point is interesting, because it highlights

---

[16] These form the 'Ramsey Collection', held by the Hillman Library of the University of Pittsburgh.

[17] See Ramsey (1991b).

[18] In Ramsey (1931), pp. 199-211. These are the notes called: 'Reasonable Degree of Belief', 'Statistics' and 'Chance', all reprinted in (1990a), pp. 97-109.

Ramsey's attitude towards frequentism, which he, far from considering a viable interpretation of probability, deems inadequate. As Ramsey puts it:

> There is, for instance, no empirically established fact of the form 'In $n$ consecutive throws the number of heads lies between $n / 2 \pm \varepsilon (n)$'. On the contrary we have good reason to believe that any such law would be broken if we took enough instances of it.
>
> Nor is there any fact established empirically about infinite series of throws; this formulation is only adopted to avoid contradiction by experience; and what no experience can contradict, none can confirm, let alone establish (*ibid.*, p. 104).

Incidentally, this passage also illustrates why – as we saw – in 'Truth and Probability' Ramsey claims not to believe in frequentism.

To Campbell's frequentist view, Ramsey opposes a notion of chance ultimately based on degrees of belief. He defines it as follows:

> Chances are degrees of belief within a certain system of beliefs and degrees of belief; not those of any actual person, but in a simplified system to which those of actual people, especially the speaker, in part approximate. [...]
>
> This system of beliefs consists, firstly, of natural laws, which are in it believed for certain, although, of course, people are not really quite certain of them (*ibidem*).

In addition, the system will contain statements of the form: 'when knowing $\psi x$ and nothing else relevant, always expect $\varphi x$ with degree of belief $p$ (what is or is not relevant is also specified in the system)' (*ibidem*). Such statements together with the laws

> form a deductive system according to the rules of probability, and the actual beliefs of a user of the system should approximate to those deduced from a combination of the system and the particular knowledge of fact possessed by the user, this last being (inexactly) taken as certain (*ibidem*).

The first characteristic of chance is that of being referred to systems containing laws. If a system related to the given phenomena is not available, then chances are also unavailable.

Ramsey stresses that chances 'must not be confounded with frequencies', for the frequencies actually observed do not necessarily coincide with them. So the chance of a coin falling heads is 1/2 even if the observed frequency of its falling heads yesterday was different. Unlike frequencies, chances can be said to be 'objective' in two ways. Firstly, to say that a system includes a chance value referred to a phenomenon, means that the system itself cannot be modified so as to include a pair of deterministic laws, ruling the occurrence and non-occurrence of the same phenomenon. As explicitly admitted by Ramsey, this characterization of objective chance is reminiscent of Poincaré's treatment of the matter, and typically applies 'when small causes produce large effects' (*ibid.*, p. 106). Secondly, chances can be said to be objective 'in that everyone agrees about them, as opposed e.g. to odds on horses' (*ibidem*).

Having so defined the notion of chance taken in general, as applied to phenomena like 'games of chance, births, deaths, and all sorts of correlation coefficients', Ramsey proceeds to its possible application to physics. He says:

> Probability in physics means chance as here explained, with possibly some added complexity because we are concerned with a 'theory' in Campbell's sense, not merely an ordinary system which is a generalization of Campbell's 'law' (*ibid.*, p. 107).

He then goes on to say that a clarification of the notion of 'theory' is needed in order to fully understand chance. As a matter of fact, in his celebrated note 'Theories' – written in 1929, one year later than 'Chance' – Ramsey builds on Campbell's notion.

Without going into details of Ramsey's or even Campbell's views of theories, it is worthwhile recollecting Campbell's distinction between laws and theories. He saw laws and theories as different constructs, that play different roles in science. While 'laws are propositions asserting relations which can be established by experiment or observation' (Campbell 1920, p. 38), theories 'cannot be proved [...] by direct experiment' (*ibid*, p.130). Furthermore, laws and theories play different roles within science. Laws describe and organize our experience by referring particular facts to general principles which usually express some form of association. Laws also allow the derivation and prediction of particular facts. On the other hand, theories allow the derivation and explanation of laws. They also make possible the expansion of scientific knowledge through the prediction of new laws. Indeed, for Campbell the ability to predict new laws does not only constitute a test for the validity of theories, it represents the main criterion for their acceptance. Those theories which gain 'universal assent' in the long run become part of science and are taken as true.

This characterization of 'true' theories given by Campbell is behind

Ramsey's notion of 'probability in physics', together with the position that in 'General Propositions and Causality' Ramsey calls 'Peirce's notion of truth as what everyone will believe in the end': a notion that applies specifically 'to the "true scientific system"' (*ibid.*, p. 161). Ramsey's notion of 'probability in physics' fits the framework of a pragmatic characterization of science as a system of beliefs shared by everyone. Probability in physics is part of our theories and represents some kind of 'ultimate chance', construed in the sense that 'there is no law [...] known or unknown, which determines the future from the past' (*ibid.*, p. 106). The supposition that such ultimate chances obtain, makes reference to 'a sort of best possible system in which they have these chances' (*ibid.*, p. 107). In other words, ultimate chance is rooted in a system which is 'strong', in the sense that it receives a good deal of weight from experience. Ultimately, reference is made to the 'true' system of science, as specified above, which can be seen as 'uniquely determined', i.e. that 'long enough investigation will lead us all to it' (*ibid.*, p. 161). The crucial role played by experimentation in this connection should not pass unnoticed, as it is through experiments that the weight of our probabilities can be increased. Only evidence coming from experiments allows us to talk about ultimate chance: 'we suppose chance to be ultimate if we see no hope of replacing it by law if we knew enough facts' says Ramsey, adding that 'there is no reason to suppose it is not ultimate' (*ibid.*, p. 159). While reaffirming Ramsey's overarching empiricism, this also illustrates his attitude towards indeterminism.

Exactly like causal laws, Ramsey takes chances to be 'variable hypotheticals', namely 'rules for judging', that 'form a system with which the speaker meets the future' (*ibid.*, p. 149). Actually, according to Ramsey all 'general propositions in the secondary system' of a theory are treated as variable hypotheticals (*ibid.*, p. 137). Since 'no proposition of the secondary system can be understood apart from the whole theory to which it belongs' (*ibidem*), chances acquire meaning only within a theory, and cannot be taken in isolation from a theoretical framework[19]. Ramsey's notion of 'objective' chance and probability in physics is obviously inextricably intertwined with his pragmatic conception of theories, truth and knowledge in general.

Ramsey's idea that within the framework of subjective probability one can make sense of an 'objective' notion of physical probability has passed almost

---

[19] For a more detailed account of Ramsey's views on theories, see Majer (1989) and (1991). Dokic and Engel (2001) contains an extended discussion of Ramsey's conception of knowledge and truth, and Sahlin (1990) offers a useful introduction to Ramsey's thought.

unnoticed. It is, instead, an important contribution to the subjective interpretation, and its application to science. As we will see, this idea finds no analogue in de Finetti's subjectivism.

## 7.3 Bruno de Finetti and exchangeability[20]

### De Finetti's radical probabilism

Bruno de Finetti was born in Innsbruck in 1906 and died in Rome in 1985. At the time Bruno was born, his father – an engineer – was director of the construction of the alpine electric railroad from Innsbruck to Fulpmes. Until the collapse of the Austro-Hungarian empire in 1918, the members of the de Finetti family were Austrian subjects, but of 'irredentist' tradition, as Bruno recollected in the autobiographical sketch 'Probability and my Life'[21]. After the sudden death of his father in 1912, Bruno and his mother, who was pregnant with his sister, moved to Trento – his mother's home town – where he attended primary and secondary school. In 1923 he entered the Polytechnic in Milan, to continue the tradition of his family, that counted many engineers. In his third year as a student, Bruno moved to the newly opened Faculty of mathematics, from which he graduated in 1927, only twenty-one years old, having already published three papers in important scientific journals. After a few years spent working at the 'Istituto Nazionale di Statistica' in Rome and subsequently at the 'Assicurazioni Generali' in Trieste, de Finetti was appointed lecturer, and subsequently professor of statistics at the University of Trieste. Then, from 1954 to 1976 he was professor of the calculus of probabilities at the University of Rome, where he lived until his death.

Bruno de Finetti was a man of strong feelings, in many ways an extremist. The fact that he grew up in an irredentist family probably underlies his radicalism in politics. As a young man, he adhered to fascism, welcoming the nationalistic character of the movement, as well as its collectivistic tendency; he opposed the liberal idea that equilibrium can be obtained through individual profit, and heralded collective economy as a way of achieving social justice. At the same time, he was a libertarian, who later in his life favoured the introduction of divorce and abortion in Italy, struggled for peace and took part

---

[20] Various passages of this section are taken from Galavotti (1989) and (2001a).
[21] See de Finetti (1982).

in the anti-militarist movement. In the Seventies, he was imprisoned with other activists of the radical party, during a pro divorce demonstration. Always very outspoken and straightforward both in his professional and everyday life, he never hesitated to criticize the ideas he did not share. De Finetti's vivid and incisive prose reflects his strong personality. His habit of inventing new expressions, as a rhetorical device to clarify his position, or making jokes of other people's attitudes, have become proverbial.

Like Ramsey, de Finetti had a natural inclination for mathematics. A letter to his mother, written in 1925 by the then 19-year-old Bruno, to plead for permission to quit engineering to study mathematics, gives an idea of his dedication to mathematics. There he writes:

> mathematics [...] is always progressing, getting richer and sharper, a lively and vital creature, in full development, for these reasons I love it, I study it, and I wish to devote my entire life to it. I wish to understand what the professors teach and I want to teach it to others, and I swear, swear that I will proceed further, because I know that I can, and I want to do it (Fulvia de Finetti 2000, p. 733).

He then resorts to some very touching expressions, to describe his highly emotional state, after attending a few lectures delivered at the newly opened Faculty of mathematics:

> Singlets, multiplets, covariant and countervariant tensors, variational calculus ... I felt like the poor boy who, standing on tiptoe, discovers an enchanted garden beyond an unsurmountable fence. [...] every formula, at least for someone who feels mathematics as I feel it, is a spark of a higher universe, that man conquers and absorbs with divine voluptuousness (*ibid.*, p. 734).

Indeed, de Finetti's contribution to the field he so enthusiastically entered, is outstanding. From his first paper, written in 1926 at the age of twenty, to his death at the age of seventy-nine, he published more than 290 works, including books, articles and reviews.

Working in the same years as Ramsey, but independently, de Finetti forged a similar view of probability as degree of belief, obeying the sole requirement of coherence. To such a definition of probability, he added the notion of exchangeability, which, combined with Bayes' rule, gives rise to the inferential methodology which is at the root of the so-called neo-Bayesianism. Surprisingly enough, despite the fact that his main technical results – including the notion of exchangeability – were found in the Twenties and Thirties, de Finetti came to be known to the scholars in the field of probability and statistics only in the Fifties, through the work of Leonard Jimmie Savage, with whom he entertained a fruitful exchange of ideas. Since then, his work is universally recognized as of crucial importance. In addition to making a substantial contribution to probability theory and statistics, de Finetti put forward an original philosophy of probability, which can be described as a blend of pragmatism, operationalism and what we would today call 'anti-realism'. De Finetti's technical results are couched in his philosophical perspective, where they acquire a peculiar meaning, which is essentially amenable to the author's firm conviction that probability cannot but be conceived in a subjective fashion.

De Finetti's philosophical position – labelled by Richard Jeffrey 'radical probabilism'[22] – reaffirms a conception of scientific knowledge as a product of human activity, ruled by (subjective) probability, rather than truth or objectivity. In the article 'Probabilismo', which he regarded as his philosophical manifesto, de Finetti traces back his own philosophy to Mach's phenomenalism. Also of some influence on de Finetti's thought were the so-called Italian pragmatists, including Giovanni Vailati, Antonio Aliotta and Mario Calderoni. The article 'Probabilismo' starts from a refusal of the notion of truth, and the related view that there are 'immutable and necessary' laws. There he writes:

> no science will permit us say: this fact will come about, it will be thus and so because it follows from a certain law, and that law is an absolute truth. Still less will it lead us to conclude skeptically: the absolute truth does not exist, and so this fact might or might not come about, it may go like this or in a totally different way, I know nothing about it. What we can say is this: I foresee that such a fact will come about, and that it will happen in such and such a way, because past experience and its scientific elaboration by human thought make this forecast seem reasonable to me (de Finetti 1931b, English edition 1969, p. 170).

Probability is precisely what makes a forecast possible. And since a forecast is always referred to a subject, being the product of his experience and convictions, the instrument we need is the subjective theory of probability. One

---

[22] See 'Introduction: Radical Probabilism', in Jeffrey (1992a), pp. 1-13; and 'De Finetti's Radical Probabilism' (Jeffrey 1993).

might say that probabilism represents for de Finetti the way out of the antithesis between absolutism and skepticism, and at its core lies the subjective notion of probability.

According to de Finetti, probability 'means degree of belief (as actually held by someone, on the ground of his whole knowledge, experience, information) regarding the truth of a *sentence*, or *event E* (a fully specified "single" event or sentence, whose truth or falsity is, for whatever reason, unknown to the person)' (de Finetti 1968, p. 45). Of this notion, the author wants to show not only that it is the only non contradictory one, but also that it covers all uses of probability in science and everyday life. This programme is implemented in two steps: firstly, an operational definition of probability is worked out, secondly, it is argued that the notion of objective probability is reducible to that of subjective probability.

A first option to work out an operational definition of probability, is in terms of betting quotients. Accordingly, the degree of probability assigned by an individual to a certain event is identified with the betting quotient at which he *would be* ready to bet a certain sum on its occurrence. The individual in question should be thought of as one in a condition to bet whatever sum against any gambler whatsoever, free to choose the betting conditions, like someone holding the bank at a gambling-casino. Probability is defined as the fair betting quotient he would attach to his bets. Like Ramsey, de Finetti regards coherence as the fundamental and unique criterion to be obeyed to avoid a sure loss, and spells out an argument to the effect that coherence is a sufficient condition for the fairness of a betting system, showing that a coherent gambling behaviour satisfies the principles of probability calculus, which can be derived from the notion of coherence itself.

In 'Sul significato soggettivo della probabilità' (de Finetti 1931a), after giving an operational definition of probability in terms of coherent betting systems, de Finetti introduces an alternative, qualitative definition of subjective probability, based on the relation of 'at least as probable as', analogous to that (independently) adopted by Jeffreys[23]. He then argues that it is not essential to embrace a quantitative notion of probability, and that, while betting quotients are apt devices for measuring and defining probability in an operational fashion, they are by no means an essential component of the notion of probability, which is in itself a primitive notion, expressing 'an individual's psychological perception' (de Finetti 1931a, English edition 1992, p. 302). The

---

[23] See Section 6.7.

idea that probability can be defined in various ways is a central feature of de Finetti's perspective, where the scheme of bets is just a convenient way of making probability understandable to the 'man in the street'. This is stressed in de Finetti's *Teoria delle probabilità*, where the betting scheme is described as a handy tool, leading to 'simple and useful insights' (de Finetti 1970, English edition 1975, vol. 1, p. 180). In addition, this book adopts another way of measuring probability, making use of scoring rules based on penalties, which is shown to be equivalent to the first method.

The autonomous value assigned to the notion of probability marks a difference between de Finetti's position and that of Ramsey and other supporters of subjectivism, like Savage. Unlike them, de Finetti does not see probability as strictly connected with utility, but as a subject characterized by an intrinsic value, to be dealt with autonomously. Commenting on Savage's work, he claims:

> The unification of the theories of probability and utility within decision theory gives the entire construction an organic and harmonious structure. However, I hesitate to follow Savage on this path; in fact such concepts, as far as I can see, have different 'cogent values': an indisputable value in the case of probability, a rather uncertain value in the case of both utility and conditions of rationality for behaviour under risk (de Finetti 1957, p. 7).

The second step of de Finetti's programme for establishing the subjective interpretation of probability consists in the reduction of objective to subjective probability. This is done by means of what is known as the 'representation theorem', obtained by de Finetti already in 1928, though its best known formulation is contained in 'La prévision: ses lois logiques, ses sources subjectives' (1937). The pivotal notion in this context is that of 'exchangeability', which corresponds to Carnap's notion of 'symmetry' and Johnson's 'permutation postulate'[24]. Summarizing de Finetti, events belonging to a sequence are *exchangeable* if the probability of $h$ successes in $n$ events is the same, for whatever permutation of the $n$ events, and for every $n$ and $h \leq n$. The representation theorem says that the probability of exchangeable events can be represented as follows. Imagine the events were probabilistically

---

[24] In his 'farewell lecture' at the University of Rome, de Finetti says that the term 'exchangeability' was suggested to him by Maurice Fréchet in 1939. Before adopting this terminology, de Finetti had made use of the term 'equivalence'. See de Finetti (1976), p. 283.

independent, with a common probability of occurrence $p$. Then the probability of a sequence $e$, with $h$ occurrences in $n$, would be $p^h (1 - p)^{n-h}$. But if the events are exchangeable, the sequence has a probability $P(e)$, representable according to de Finetti's representation theorem as a *mixture* over the $p^h (1 - p)^{n-h}$ with varying values of $p$:

$$P(e) = \int p^h (1 - p)^{n-h} \, dF(p)$$

where the distribution function $F(p)$ is unique. In the above equation two kinds of probability are involved, namely the subjective probability $P(e)$ and the 'objective' (or 'unknown') probability $p$ of the events considered. This enters into the mixture associated with the weights assigned by the function $F(p)$, which represents a probability distribution over the possible values of $p$. Assuming exchangeability, then, amounts to assuming that the events considered are equally distributed and independent, given any value of $p$.

In order to understand de Finetti's position, it is useful to start by considering how an objectivist would proceed when assessing the probability of an unknown event. An objectivist would assume an objective success probability $p$. But its value would in general remain unknown. One could give weights to the possible values of $p$, and determine the weighted average. The same applies to the probability of a sequence $e$, with $h$ successes in $n$ independent repetitions. Note that because of independence it does not matter where the successes appear. De Finetti focuses on the latter, calling *exchangeable* those sequences where the places of successes don't make a difference in probability. These need not be independent sequences. An objectivist who wanted to explain subjective probability, would say that the weighted averages are precisely the subjective probabilities. But de Finetti proceeds in the opposite direction, with his representation theorem. It says in his interpretation: starting from the subjective judgment of exchangeability, one can show that there is only one way of giving weights to the possible values of the unknown objective probabilities. According to this interpretation, objective probabilities become useless and subjective probability can do the whole job.

In the course of a comment on the notion of exchangeability, de Finetti reaffirms that the latter represents the correct way of expressing the idea that is usually conveyed by the expression 'independent events with constant but unknown probability'. If we take an urn of unknown composition, says de Finetti, the above phrase means that, relative to each of all possible compositions of the urn, the events can be seen as independent with constant probability. Then he points out that

> what is unknown is the composition of the urn, not the probability: the latter is always known and depends on the subjective opinion about the composition, which opinion is modified as new drawings are made, and observed frequencies are taken into account (de Finetti 1995, p. 214).

It should not pass unnoticed that for de Finetti subjective probability, being the expression of the feelings of the subjects evaluating it, is always definite and known: 'probability as *degree of belief* is surely known by anyone' (de Finetti 1973, p. 356).

From a philosophical point of view, the reduction of objective to subjective probability is to be seen from a pragmatic angle. It is performed in the same pragmatic spirit that inspires the operational definition of subjective probability, and complements the latter. If such a reduction is based on consideration of the role played by objective probability in statistical reasoning, it is again the role played by subjective probability in life and science that gives an operational basis for its definition. 'Probability – says de Finetti – is actually already *defined* implicitly by the role played, with respect to the decisional criterion of an individual, by the fact that he evaluates it in a certain way' (de Finetti 1963, p. 66). So much for the philosophical significance of the representation theorem within de Finetti's subjectivism.

From a more general viewpoint, the representation theorem gives applicability to subjective probability, by bridging the gap between degrees of belief and observed frequencies. Taken in connection with Bayes' rule, exchangeability provides a model of how to proceed in such a way as to allow for an interplay between the information on frequencies and degrees of belief. By showing that the adoption of Bayes' method, taken in conjunction with exchangeability, leads to a convergence between degrees of belief and frequencies, de Finetti indicates how subjective probability can be applied to statistical inference. It is for this reason that de Finetti's exchangeability can be regarded as the decisive step towards the edification of modern subjectivism.

De Finetti is convinced that the representation theorem answers Hume's problem because it justifies 'why we are also intuitively inclined to expect that frequency observed in the future will be close to frequency observed in the past' (de Finetti 1972a, p. 34). De Finetti's argument is pragmatic, being based on the role of induction: to guide inductive reasoning and behavior in a coherent way. His view is actually in full agreement with Hume's perspective: like the latter, de Finetti regards it as impossible to give a logical justification of induction, and answers the problem in a psychologistic fashion. A claim repeated by de Finetti on various occasions is that 'we must begin again from where Hume left off' (de Finetti, 1938a, English edition 1985, p. 79). By giving a mathematical argument that shows the coherence of our inductive habits, de Finetti aims at proceeding one step further along the road indicated by Hume.

## Subjective Bayesianism

According to de Finetti, statistical inference can be entirely performed by exchangeability in combination with Bayes' rule; therefore, de Finetti's probabilism is intrinsically Bayesian. If his notion of probability as degree of belief is grounded in an operational definition, probabilistic inference – taken in a subjective sense – is made to rely on Bayes' theorem. One might say that for de Finetti Bayesianism represents the crossroads where pragmatism and empiricism meet subjectivism, for he thinks that one needs to be Bayesian in order to be a subjectivist, while subjectivism is the only choice open to those embracing a pragmatist and empiricist philosophy.

Having adopted such an uncompromising form of Bayesianism, de Finetti regards the shift from prior to posterior – or, as he preferred to say, from initial to final – probabilities as the cornerstone of statistical inference. In this connection, he takes a 'radical approach' by which 'all the assumptions of an inference ought to be interpreted as an overall assignment of initial probabilities' (de Finetti 1969, p. 9)[25]. This shift is given a subjective interpretation, in the sense that going from prior to posterior assessments involves a shift from one subjective probability to another, though objective factors, like frequencies, are obviously taken into account, when available. De Finetti calls attention to the fact that, according to his interpretation, updating one's mind in view of new evidence does not mean changing opinion:

> If we reason according to Bayes' theorem we do not change opinion. We keep the same opinion and we update it to the new situation. If yesterday I said 'Today is Wednesday', today I say 'It is Thursday'. Yet I have not changed my mind, for the day following Wednesday is indeed Thursday (de Finetti 1995, p. 100).

If the idea of correcting previous opinions is completely alien to this perspective, so is the notion of a self-correcting procedure, which, as we have

---

[25] See also de Finetti and Savage (1962).

seen, occupies a central place within the perspective of other authors, such as Hans Reichenbach. De Finetti's attitude is grounded in the conviction that there are no 'correct' and 'rational' probability assignments: 'the subjective theory [...] does not contend that the opinions about probability are uniquely determined and justifiable. Probability does not correspond to a self-proclaimed "rational" belief, but to the effective personal belief of anyone' (de Finetti 1951a, p. 218). His attitude in this connection is in sharp contrast with the logicism of Rudolf Carnap and Harold Jeffreys, who believe in 'correct' probability evaluations.

De Finetti's subjective Bayesianism is intransigent, even dogmatic: not only is subjective Bayesianism the sole acceptable way of addressing probabilistic inference and the whole of statistical methodology, but it makes any form of 'objectivism' look silly. As de Finetti puts it:

> The whole of subjective statistics is based on this simple theorem of probability calculus [Bayes' theorem]. Consequently, subjective statistics has a very simple and general foundation. Moreover, being grounded only on the basic axioms of probability, subjective statistics does not depend on those definitions of probability that would narrow its range of application (like, for instance, the definitions based on the idea of equally probable events). Nor – once one endorses this view – is there any need to resort to empirical formulae, in order to characterize inductive reasoning. Objectivist statisticians, on the contrary, make extensive use of empirical formulae. The need to do so stems only from their refusal to admit the use of initial probability. [...] They reject the use of initial probability because they reject the idea of a probability that depends on the state of information. However, by doing so they distort everything: not only do they make probability an objective entity [...] they even make it a theological entity: they claim that 'true' probability exists, outside us, independently of a person's judgment (de Finetti 1995, p. 99).

This passage highlights another feature of de Finetti's position, namely his rejection of objective probability, which is deemed not only useless, but meaningless, like all metaphysical notions. Throughout his life, de Finetti held that 'probability does not exist'. This claim, which appears in capital letters in the 'Preface' to the English edition of *Teoria delle probabilità*, is the *leit-motiv* of his production. 'Objective probability never exists' he writes in 'Il significato soggettivo della probabilità' (de Finetti 1931a, English edition 1992,

p. 295), and almost fifty years later he opens the article 'Probabilità' in the *Enciclopedia Einaudi* with the words: 'is it true that probability "exists"? What could it be? I would say no, it does not exist' (de Finetti 1980, p. 1146). Such aversion to the ascription of an objective meaning to probability is inspired by the desire to keep the notion of probability free from metaphysical 'contaminations'[26].

De Finetti's refusal of objective probability goes hand in hand with his lack of consideration for the notions of 'chance' and 'physical probability'. In the conviction that science is just a continuation of everyday life, de Finetti never paid much attention to the use made of probability in science, and stuck to the conviction that subjective probability is all that is needed. Only the volume *Filosofia della probabilità* – containing the text of a lecture course given by de Finetti in 1979 – includes a few remarks to the effect that probability distributions belonging to statistical mechanics can be taken as more solid grounds for subjective opinions (de Finetti 1995, p. 117). Such remarks may be taken to suggest that late in his life de Finetti must have entertained the idea that some special meaning can be attached to those probability assignments, which derive from accepted scientific theories. Unlike Ramsey, however, de Finetti did not feel the need to include in his theory a notion of probability specifically devised for application in science. This feature of his theory, together with his criticism of testing procedures and his refusal of other 'objective' notions, such as that of randomness, is likely to be responsible for the fact that subjectivism has never been taken very seriously by scientists[27].

Having refused the notion of 'objective' probability and denied that there are 'correct' probability assignments, the radical subjectivist de Finetti still faces the problem of objectivity of probability evaluations. He believes that the process by which probability judgments are obtained is more complex than is supposed by all of the other interpretations of probability, which define probability on the basis of a unique criterion. On the contrary, de Finetti regards the fact that the subjective viewpoint makes a sharp distinction between the definition and the evaluation of probability as a decisive advantage. While subjectivists do not mix these things up, upholders of the other interpretations confuse them: they look for a unique criterion – be it frequency, or symmetry – and use it as grounds for both the definition and the evaluation of probability. In so doing, they embrace a 'rigid' attitude towards

---

[26] See Galavotti (1989) for an exposition of the anti-metaphysical and anti-realist basis of de Finetti's subjectivism.

[27] This is argued also in Galavotti (1995-96) and (1996).

probability, an attitude which consists 'in defining (by whichever way and whichever concept) the probability of an event, thus univocally determining a function' (de Finetti 1933, English edition 1992, p. 384). Instead, subjectivists adopt an 'elastic' approach, which

> consists in demonstrating that all functions $f$ have the necessary and sufficient properties to represent not-intrinsically contradictory probability evaluation, (here also defined according to whichever concept and by whichever road), leaving then to a second (extra-mathematical) phase the discussion and the analysis of motives and criteria for the choice of a particular one amongst all these possible evaluations (*ibidem*).

Subjectivists regard all coherent functions as admissible; far from committing themselves to a single rule or method, they take the choice of one particular function as the result of a complex and largely context-dependent procedure, which necessarily involves subjective elements.

The explicit recognition of the role played by subjective elements within the complex process of the formation of probability judgments is for de Finetti a prerequisite for the appraisal of objective elements. So, he maintains that subjective elements in no way 'destroy the objective elements nor put them aside, but bring forth the implications that originate only after the conjunction of both objective and subjective elements at our disposal' (de Finetti 1973, p. 366). Starting from the assumption that a probability evaluation expresses one's degree of belief, Bayesian subjectivism requires that objective elements also be taken into account. De Finetti is very clear on this point:

> Every probability evaluation essentially depends on two components: (1) the objective component, consisting of the evidence of known data and facts; and (2) the subjective component, consisting of the opinion concerning unknown facts based on known evidence (de Finetti 1974, p. 7).

However, he warns that the objective component of probability judgments, namely factual evidence, is in many ways context-dependent: evidence must be collected carefully and skilfully, its exploitation depends on judging what elements are relevant to the problem under consideration, and can be useful to the evaluation of related probabilities. In addition, the collection and exploitation of evidence depends on economic considerations varying in practical cases. So, one can say that the collection and exploitation of factual evidence involves subjective elements of various sorts. Equally subjective is the decision on how to let objective elements influence belief. Typically, one relies on information regarding frequencies. When a considerable amount of information about frequencies is available, it influences probability assignments through the adoption of exchangeability.

If information on frequencies is scant, one faces the problem of how to obtain good probability evaluations. De Finetti addressed this problem in a number of writings, partly fruit of his cooperation with Savage. The approach adopted is based on penalty methods, of the kind of the well known 'Brier's rule'[28]. Scoring rules are devised so as to oblige those who make probability evaluations to be as accurate as they can and, if they have to compete with others, to be honest. Such rules play a twofold role within de Finetti's approach. In the first place, they offer a suitable tool for an operational definition of probability, which is in fact adopted by de Finetti in his late works. In addition, these rules offer a method for improving probability evaluations made both by a single person and by several people, because they can be employed as methods for exercising 'self-control', as well as a 'comparative control' over probability evaluations (de Finetti 1980, p. 1151). De Finetti gives these methods a straightforward interpretation, in tune with his subjectivism. In this connection, he writes:

> The objectivists, who reject the notion of personal probability because of the lack of verifiable consequences of any evaluation of it, are faced with the question of admitting the value of such a 'measure of success' as an element sufficient to soften their fore-judgment. The subjectivists, who maintain that a probability evaluation, being a measure of someone's beliefs, is not susceptible of being proved or disproved by the facts, are faced with the problem of accepting some significance of the same 'measure of success' as a measure of the 'goodness of the evaluation'. [...] as for myself, though maintaining the subjectivist idea that no fact can prove or disprove belief, I find no difficulty in admitting that any form of comparison between probability evaluations (of myself, of other people) and actual events may be an element influencing my further judgment, of the same status as any other kind of information (de Finetti 1962, p. 360).

De Finetti's work on scoring rules is in tune with a widespread attitude among

---

[28] De Finetti's adoption of this method is discussed in more detail in Galavotti (2001a).

Bayesian statisticians, an attitude that has given rise to a vast literature on 'well-calibrated' estimation methods.

In conclusion, de Finetti regards the evaluation of probability as a complex procedure, resulting from the concurrence of a myriad factors. Starting from the recognition that probability is subjective, and that there is no unique, 'rational' way of assessing it, one should take into account a whole array of elements that can influence probability evaluations, because it is only through recognition of such elements that the estimation of probability can be enhanced. What should be avoided at all costs is the identification of objectivity with objectivism, for this would be a mistake of the worst kind. It is precisely this kind of mistake that, according to de Finetti, tarnishes all the other interpretations of probability.

## Criticism of other interpretations of probability

The main target of de Finetti's criticism is frequentism, which he accuses of being a 'conceptual distortion'. In one of the lectures collected in the volume *Filosofia della probabilità*, bearing the title 'Superstition and frequentism', we read that

> There is no worse conceptual distortion than that owing to which, starting from the premise that any sequence can occur, one defines probability in terms of a property (that of exhibiting a certain frequency) pertaining only to a portion of all sequences. [...] when we define probability in terms of frequency, we define it thoughtlessly. The only objective thing is the set of all possible sequences, but it does not say anything concerning their probability. The probability of sequences can only be the feeling we had before and which characterized our expectation. Here we have a perversion of language, logic and common sense. Such a logical mistake is unacceptable, because the set of all possible sequences (which is logically determined) cannot be confused with probability (which, on the contrary, is subjective) (de Finetti 1995, pp. 140-1).

The alleged contradiction undermining frequentism arises from the fact that the link between probability and frequency is in no case to be taken as given, being rather a matter for argument. Indeed, 'one of the central tasks of probability calculus is that of *demonstrating* the relation between probability and frequency, but this becomes impossible if such a relation is postulated' (de Finetti 1938b, p. 8). Since it takes as an assumption what, instead, ought to be demonstrated, frequentism is ill-founded: 'from the frequentist standpoint, frequency converges to probability by definition. But unfortunately there is no guarantee for it. Frequentism originates from the mystification which pretends that probabilistic laws are certain' (de Finetti 1995, p. 132).

De Finetti is also critical of the basic notions underpinning frequentism, such as von Mises' notion of collective. As to the infinite character of collectives, de Finetti observes that 'the idea of an infinite series of experiments' is simply 'meaningless' (*ibid.*, p. 130). In fact, 'even supposing that the universe will last billions of times longer than one thinks it will last, it cannot but have a finite duration' (*ibid.*, p. 124). In a similar fashion, he objects to the notion of randomness, and claims that that of an 'irregular sequence' is a '*pseudo-notion*', adding that

> pseudo-notions are always a great obstacle in dealing with whatever issue. One should avoid such conceptual outgrowths, originating from superstitions, like that according to which the long awaited numbers of a lottery are more likely to be drawn, and other suchlike cabalistic beliefs (*ibid.*, p. 134).

De Finetti welcomes Wald's criticism of von Mises' notion of randomness, which he takes as a sign of decline of the consensus around the frequency interpretation of probability[29]. Not surprisingly for the uncompromising subjectivist he is, de Finetti ascribes little importance to the notion of randomness.

De Finetti also moves various objections to Reichenbach's version of frequentism. In particular, he rejects the notion of homogeneity of the reference class. The latter for de Finetti can only be a matter for subjective evaluation, leading to a decision to consider certain elements members of the same sequence. To be sure, when we are in a position to judge that certain events show enough similarities, to allow us to take them as belonging to a given sequence, we are also in a position to make good use of frequencies, but one has to be clear about the fact that, if under appropriate circumstances frequency can provide a criterion for practical evaluation of probability, it should never be confused with probability itself.

Remarkably, there are also aspects of Reichenbach's position that de Finetti finds congenial. He agrees with Reichenbach's tenet that theory of probability

---

[29] See de Finetti (1951b), containing a comment on Wald's work, and (1936) for de Finetti's criticism of von Mises.

should address the problem of the justification of induction. Unlike von Mises, who, according to de Finetti, puts forward a 'fragmentary' viewpoint, Reichenbach rightly 'argues for the logical inconsistency and scientific inadequacy of a theory of probability that in face of the problem of inductive reasoning does not attempt a solution, but rather "closes its eyes prudishly"' (de Finetti 1941, pp. 126-7). Like Reichenbach, de Finetti attaches great importance to the justification of induction, and their arguments – albeit framed within different, even opposite, philosophies of probability – exhibit some analogies. For both of them the notion of induction is not primitive, but derives from that of probability. Furthermore, they both regard Bayes' rule as the pre-eminent method of induction and appeal to pragmatic considerations to justify induction. Where Reichenbach's and de Finetti's positions are irreconcilable is in connection with the idea of a self-correcting procedure, which occupies a pivotal role within Reichenbach's epistemology, and – as we have seen – is deemed meaningless by de Finetti.

De Finetti's criticism of the logical interpretation of probability moves at least partly along similar lines, for he charges its supporters with making the same mistake made by frequentists, defining probability on the basis of something external to it, instead of taking it as a primitive notion. In doing so, they all adopt a 'rigid' attitude toward probability, and assume that there are objective reasons supporting the choice of a particular probability function among all admissible ones. Such reasons are for logicists based on consideration of the logical properties of propositions. For de Finetti, this amounts to building the notion of probability 'on "the structure of nothing", or, to put it differently, on the attribution of a realistic content to logical-mathematical expressions which are purely formal or symbolic' (de Finetti 1938a, English edition 1985, p. 86). This criticism, addressed in the Thirties to the 'Cambridge probabilists' – namely John Maynard Keynes and Harold Jeffreys – reappears thirty years later in a comment on Carnap's work. There de Finetti observes that Carnap 'would like to rely on the structure or syntax of propositions' to give a logical foundation to probability judgments, and objects that 'formal reasons regarding words, or symbols, or syntactical conventions' have no relevance for probability evaluations (de Finetti 1967, p. 205). He further points out that, by stressing the importance of the symmetrical function $c^*$, Carnap comes very close to subjectivism, but between their views there remains a conceptual gap. Furthermore, commenting on Carnap's 'The Aim of Inductive Logic', de Finetti claims to be in complete agreement with Carnap, 'if his "credibility function" is interpreted simply as subjective personal probability' (de Finetti 1972b, p. 183). As we have seen, this does not seem to be the case, since Carnap's notion of probability is grounded in a concept of rationality which is not amenable to a pragmatical characterization.

It is noteworthy that for de Finetti those embracing the logical interpretation are right in claiming that probability is a property of propositions. At various places in his writings he claims that when he speaks of 'events', he really means 'logical entities'. In addition, he regards the theory of probability as a 'logic of uncertainty'. However, de Finetti – like Ramsey – looks at logic in a psychologistic fashion, and assigns it the purpose of teaching 'the coherence of thought with itself' (de Finetti 1930, 1981, p. 261). Logicists, on the contrary, believe that logic can inspire an objectivistic attitude towards probability, and constrain probability evaluation in what he calls a 'rigid' fashion.

## Indeterminism

The kind of epistemic conception of probability represented by de Finetti's subjectivism could not be more distant from Laplace's perspective. Needless to say, de Finetti also regards the classical interpretation as an expression of the 'rigid' way of looking at probability. Moreover, he claims that 'the belief that the *a priori* probabilities are distributed uniformly is a well defined opinion and is just as specific as the belief that these probabilities are distributed in any other perfectly specified manner' (de Finetti 1951a, p. 222). Assumptions like those of independence and exchangeability are likewise a matter for judgment.

De Finetti's subjectivism is in sharp contrast with Laplace in a further respect, namely in connection with determinism. Unlike Laplace, de Finetti is not a determinist. The problem of determinism is addressed in 'Le leggi differenziali e la rinuncia al determinismo' (1931), where he contrasts two positions:

> 1) necessary and unchangeable laws exist: natural phenomena are determined by their antecedents with absolute accuracy and certainty;
> 2) real laws do not exist as such; forecasts cannot be certain, but only more or less, and perhaps immensely, likely or probable; the so called natural laws are no more than the expression of statistical regularities (de Finetti 1931c, English edition 1992, p. 323).

De Finetti argues that the contrast between these viewpoints is more philosophical than physical, and more subjective than objective. Nevertheless, science points to a way out of such an antithesis, because science does not rely at all on the principle of causality, which therefore appears useless, whereas it makes extensive use of probabilistic forecasting methods. De Finetti continues by claiming that, in the light of modern science, we have to admit that events are not determined with certainty, and this admission leads to a denial of determinism, and concludes that

> By rejecting determinism we must accept completely the second of the two propositions that I have stated: then, forecasts will no more be certain, but only more or less probable. One might have a probability so large as to be entitled to name it practical certainty, but this does not change the fact that it is simply a probability.
>
> The essential novelty in the scientific method would then be the substitution of logic by probability theory; instead of rationalistic science, where certainty is deduced from certainty, there would be a probabilistic science, where the probable is deduced from the probable. It is not prejudicially necessary to renounce determinism for setting up science on these bases; we may confess not to be able to foresee an event without saying that forecasting is by itself impossible. We may then develop science on this line as: 1) acceptable to everybody, either renouncing or not renouncing determinism; 2) independent from the causality principle, which then becomes totally useless. From the acknowledgement of it being useless, to abandon it, there is, psychologically speaking, a short step; however this concerns philosophy, not science (*ibid.*, p. 324).

We find in this paper, published in the same year as 'Probabilismo', the expression of de Finetti's pragmatism, combined with a genuinely empiricist attitude. Both determinism and indeterminism are unacceptable, when taken as physical, or even metaphysical, hypotheses. They can at best be useful ways of describing certain facts, provided that they are not taken in the above sense.

This view is reaffirmed by de Finetti in his 'farewell lecture', delivered at the University of Rome before his retirement in 1976, where he says that

> the alternative [between determinism and indeterminism] is undecidable and (I should like to say) illusory. These are metaphysical diatribes over 'things in themselves'; science is concerned with what 'appears to us', and it is not strange that, in order to study these phenomena it may in some cases seem more useful to imagine them from this or that standpoint (de Finetti 1976, p. 299).

## 7.4 Some recent trends

Subjectivism is currently popular with statisticians and philosophers of science. The fortunes of subjectivism are to a certain extent tied up with those of Bayesianism, an essential ingredient of the subjective approach, although the class of Bayesians is much broader than that of subjectivists, and there are Bayesians like Hans Reichenbach and Harold Jeffreys who embrace an objectivist notion of probability. More recently, an objective version of Bayesianism was put forward by Edwin T. Jaynes, who fostered an objective way of assessing priors, based on the 'principle of maximum entropy'[30].

It would be exceedingly difficult to attempt a survey of recent Bayesian literature, if we have to credit Irving John Good, who claimed to distinguish 46,656 different varieties of Bayesians![31] In what follows attention will focus only on the work of two authors, both of whom have taken inspiration from Bruno de Finetti to develop original epistemological viewpoints, deeply imbued with probability.

### Richard Jeffrey's radical probabilism[32]

Richard Jeffrey (1926-2002) studied under Carnap in Chicago in the years 1946-51, and in 1957 obtained a doctorate from Princeton, where he then spent most of his life, as professor of philosophy. After being initiated to philosophy of science and logic by Carnap[33], Jeffrey was strongly influenced by what he labelled de Finetti's 'radical probabilism', of which he declared himself a follower, and which he regarded as a radicalization of Carnap's position.

Jeffrey embraces a form of Bayesianism which revolves around the fundamental tenet that the entire edifice of human knowledge rests on probability judgments, not on certainties. In the 'Introduction' to the collection of essays published under the title *Probability and the Art of Judgment*, Jeffrey outlines different ways of being a Bayesian:

> Broadly speaking, a Bayesian is a probabilist, a person who sees making up the mind as a matter of either adopting an assignment of judgmental probabilities or adopting certain features of such an

---

[30] Jaynes work is too technical to be reviewed here. The reader is addressed to Jaynes (1983) and (2003).

[31] See '46,656 Varieties of Bayesians' in Good (1983).

[32] This section is partly taken from Galavotti (1996).

[33] Some important manuscripts by Carnap were published posthumously by Jeffrey in Carnap and Jeffrey, eds. (1971) and Jeffrey, ed. (1980).

assignment, e.g., the feature of assigning higher conditional probability to 5-year survival on a diagnosis of ductal cell carcinoma than on a diagnosis of islet cell carcinoma (Jeffrey 1992a, p. 2).

In a narrower sense, a Bayesian is taken to be one who sees 'conditioning (or "conditionalization") as the only rational way to change the mind' (*ibidem*); a position with which Jeffrey claims disagreement. Furthermore, Jeffrey baptizes as a 'rationalist Bayesian' a Bayesian in this narrower sense, who thinks that 'there exists a (*logical, a priori*) probability distribution that would define the state of mind of a perfect intelligence, innocent of all experience' (*ibidem*). Rationalistic Bayesianism is taken to be a fairly broad category, that includes not only the fathers of the logical interpretation of probability, namely Johnson, Keynes and Carnap, but also Laplace and the Reverend Bayes himself. The hallmark of this form of Bayesianism is the conviction that judgments have two components, one purely rational and one purely empirical, and that they can be analysed in terms of them. The rationalistic component is represented by the assumption of an 'ignorance prior', like the Laplacean equiprobability distribution, or the 'symmetric' distribution characterizing Carnap's $c^*$ function, while the empirical component is given by experience. Jeffrey's radical probabilism also rejects this view, to embrace a broader perspective, depicted as anti-rationalist, nonfoundational and pragmatist.

In the first place, Jeffrey rejects as an 'empiricist myth' the idea that experiential data can be grounded on a phenomenistic base, as well as the related claim that this kind of information forms the content of so-called 'observation sentences', that can qualify as true or false. Such an assumption – typically held by logical empiricists, including Carnap – combined with a Bayesian standpoint leads to a view according to which conditioning is based on experimental evidence taken as certain. Within Jeffrey's radical probabilism, this view is replaced by the tenet that probabilities need not be based on certainties. On the contrary, 'it can be probabilities all the way down, to the roots' (*ibid.*, p. 11). It is in this connection that radical probabilism is deemed by Jeffrey a 'nonfoundational methodology' (*ibid.*, p. 68): a methodology that does not ground knowledge on certainty.

Jeffrey traces his own position back to that of the fathers of the subjective interpretation of probability, namely Ramsey and de Finetti, while distancing himself from Carnap. The latter's idea that one can trace a separation between an empirical and a rational component of our judgments is rejected by radical probabilism, which, by way of denying this, qualifies as 'anti-rationalist'. Radical probabilism also rebuts one of the basic tenets of Carnap's programme

– that conditioning starts from the assumption of some *a priori* distribution that can be justified on logical grounds. In Jeffrey's words:

> Carnap's idea of an 'ignorance' prior cumulatively modified by growth of one's sentential data base is replaced by a pragmatical view of priors as carriers of current judgment, and of rational updating in the light of experience as a congeries of skills (*ibid.*, p. 12).

Jeffrey's departure from Carnap's programme goes in the direction of a stronger pragmatism, according to which judgments result from a multitude of factors, some of which are strictly pertinent to the evaluating subject, while others are essentially social products. In the process leading to the formulation of probability judgments, personal intuition and trained expertise mingle with a whole set of methods, theories and skills shared by the scientific community. As Jeffrey puts it:

> Modes of judgments (probabilizing, etc.) and attendant standards of rationality are cultural artifacts, bodies of practice modified by discovery or invention of broad features seen as grounds to stand on. It is ourselves or our fellows to whom we might wish to justify particular judgments (Jeffrey 1992b, p. 203).

As suggested by the above passage, Jeffrey's overarching pragmatism does not contemplate a notion of rationality beyond the sphere of human knowledge and conduct, to guide our judgment. While regarding this as the fundamental lesson that Bayesians should learn, Jeffrey warns that radical probabilism qualifies as a programme whose applicability to everyday practice cannot be canonised once and for all, but depends on context and topic dependent techniques forming 'an art of judgment going beyond honest diligence' (*ibid.*, p. 204). In conclusion:

> Shy of the pragmatism that sees judgment as topical technique, Bayesian manuals have envisaged a never-never land of seamless precision in place of our real patchwork of partial probabilistic judgments. A user's manual for radical probabilism remains to be written (*ibidem*).

Jeffrey's radical probabilism qualifies as a flexible and tolerant position, a 'Bayesianism with a human face'[34], apt to account for uncertain evidence and

---

[34] See Chapter 5 of Jeffrey (1992a).

imperfect information. At its core lies a dynamical theory of decision, called *probability kinematics*, spelled out in Jeffrey's pioneering *The Logic of Decision* (1965, 1983), which opened a new field of enquiry. Its backbone is the so-called 'Jeffrey's rule' of conditionalization, closely resembling Donkin's method described at the beginning of this chapter. Jeffrey' conditionalization is a generalization of traditional Bayesian conditioning. Its insight is conveyed by the following account, due to Brian Skyrms:

> Jeffrey's basic model assumes a finite partition each of whose members has positive prior probability. A probability, say $pr_2$, is said to come from another $pr_1$ by probability kinematics on this partition just in case the final probabilities conditional on members of the partition, where defined, remain the same as the initial probabilities conditional on members of the partition. Conditioning on a member of the partition is the special case of probability kinematics in which that member gets final probability of one. Jeffrey had in mind a model in which one could approximate certain evidence without being forced to regard learning as learning for certain (Skyrms 1996, p. 287).

The partition in question is a set of propositions whose probability is being assessed. The case in which one conditions on a member of the partition, which gets final probability of one, corresponds to classical Bayesian conditioning. As pointed out by Skyrms in the above quoted passage, the basic idea here is to give a method for updating probability judgments in the presence of new information, which is itself uncertain.

The dynamical dimension characterizing diachronic conditionalization calls for an equally dynamical notion of coherence. In recent years, dynamic coherence, or diachronic Dutch book arguments, have attracted the attention of a number of authors, and inspired a large amount of literature, whose level of technicality falls beyond the limits of by the present account[35]. It should not pass unnoticed that the dynamical dimension of conditionalization is also linked to the problem of the value of knowledge. As we saw, this problem was noticed and addressed (albeit in different terms) by Keynes and Ramsey. As to the analysis of this notion, authors of Bayesian inspiration have made

---

[35] Two articles dealing specifically with Jeffrey's conditionalization and the problem of its justification in terms of dynamic coherence are Howson (1996) and Skyrms (1996). More is to be found in Skyrms (1990), Howson (1989, 1993) and Earman (1992). Jeffrey's latest accomplishments are contained in his last book, published posthumously: *Subjective Probability: The Real Thing* (2004).

tremendous advances, starting from the pioneering studies of Leonard Jimmie Savage and John Irving Good, to the more recent work of Brian Skyrms and myriad other authors[36].

To conclude this brief outline of Jeffrey's probabilism, it is worth mentioning that it is meant to accommodate some notion of 'objective probability'. In this connection, Jeffrey disagrees with de Finetti's rejection of notions like chance, objective probability, randomness and the like, and comes closer to Ramsey's perspective, although he does not place the same emphasis on the need to ground objective probability on a suitable view of scientific theories. In an attempt at gaining subjective Bayesianism wide applicability, within his radical probabilism Jeffrey makes room for a 'non-frequentist objectivism' that admits of the notion of physical probability[37]. In this vein, Jeffrey's article 'De Finetti's Radical Probabilism' (1993) points to a way, meant to be in tune with de Finetti's subjectivism, of interpreting probabilities encountered in quantum mechanics in terms of degrees of belief[38]. This attitude is in line with a tendency to open subjectivism to objective probability and chance, which has recently grown stronger.

In this connection, it is worth mentioning the work of David Lewis (1980), who in 'A Subjectivist's Guide to Objective Chance' held the conviction that 'along with subjective credence we should believe also in objective chance', because 'the practice and the analysis of science require both concepts' (Lewis 1980, p. 263). In order to bridge these two concepts Lewis adopted a principle known as the 'Principal principle', which, as Howson and Urbach put it, states that

> if the objective, physical probability of a random event (in the sense of its limiting relative-frequency in an infinite series of trials) were known to be $r$, and if no other relevant information were available, then the appropriate subjective degree of belief that the event will occur on any particular trial would also be $r$. If the event in question is $a$, the Principal Principle says that $P[a_t \mid P^*(a) = r] = r$, where $a_t$ is a statement describing the occurrence of the event on a particular trial, $P^*(a)$ is the objective probability, and $P$ is a subjective probability function (Howson and Urbach 1989, 1993, p. 240).

This principle, which provoked a debate that is much too wide to be outlined in

---

[36] See Savage (1954); Good (1967), and Skyrms (1984) and (1990).
[37] See Jeffrey's article 'Mises Redux', reprinted as Chapter 11 of Jeffrey (1992a).
[38] See also Jeffrey (1996).

these pages, seems to capture a fundamental intuition of the Bayesian approach, while linking subjective and objective probabilities[39].

## Patrick Suppes' probabilistic empiricism[40]

A flexible form of Bayesianism has also been put forward by Patrick Suppes, whose ideas on propensity and indeterminism were highlighted at the end of Chapter 5. Born in 1922, Suppes obtained a doctorate from Columbia under the supervision of Ernest Nagel. He spent a long and very productive career at Stanford University, where he is still actively engaged in research. A distinctive trait of Suppes's personality lies with his being, besides a philosopher, a scientist in the fields of both psychology and theoretical physics. Such a double militancy won him the reputation of 'scientific philosopher'[41], and exercised a direct bearing on his philosophy of science, marked by a deeply pragmatical flavour. In a pragmatist spirit, Suppes regards philosophy of science as aimed at understanding science taken as a perpetual problem-solving activity, rather than at developing ubiquitous philosophical doctrines.

Suppes fosters a probabilistic version of empiricism, calling attention to the centrality of the notion of probability within philosophy of science and epistemology. In a book emphatically called *Probabilistic Metaphysics* (1984), Suppes writes that

it is probabilistic rather than merely logical concepts that provide a rich enough framework to justify both our ordinary ways of thinking about the world and our scientific methods of investigation (Suppes 1984a, p. 2).

One should therefore embrace a 'probabilistic empiricism', namely a probabilistic approach to science and philosophy intended to supersede the 'neotraditional metaphysics', centred on determinism, in the conviction that 'certainty of knowledge – either in the sense of psychological immediacy, in the sense of logical truth, or in the sense of complete precision of measurement – is unachievable' (*ibid.*, p. 10). Among the arguments put

---

[39] For a discussion of the Principal principle see Earman (1992) and Howson and Urbach (1989, 1993).
[40] Some passages of this section are taken from Galavotti (1994).
[41] *Patrick Suppes: Scientific Philosopher* is the title of the three volumes published to celebrate his seventieth birthday: see Humphreys, ed. (1994).

forward by Suppes to support this claim, one of the most convincing is rooted in imprecision of measurement, which does not come only in connection with human or instrumental errors, but arises in a more substantial way from certain developments of last century's physics, like Heisenberg's principle of uncertainty. Uncertainty then pervades not only the level of experimentation, but is also encountered at the level of physical theories. Recognition of uncertainty 'at the most fundamental level of theoretical and methodological analysis' (*ibid.*, p. 99) leads directly to probability, because probabilistic methods provide 'a natural way' of working out the form of empiricism advocated by Suppes. Embedded in the latter, we find the tenet that a probabilistic character pertains not only to the basic laws of natural phenomena, but also to causality and rationality. Suppes developed probabilistic theories of both causality and rationality, which cannot be described in detail[42].

Suppes' probabilistic empiricism leaves the door open to the idea that there are random phenomena in nature. But as it was pointed out at the end of Chapter 5, Suppes does not side with either of the alternatives between determinism and indeterminism: he believes that the opposition between them should be abandoned in favour of an account of phenomena in terms of complexity and/or instability. Determinism, like the principle of universal causation, the ideal of certainty, and the idea that in principle human knowledge can be completed, are deemed 'chimeras' of rationalism, representing a hindrance to a philosophy meant not to lose sight of science. Freed from the chimeras of rationalism, probabilistic empiricism can be developed along pragmatic lines leading to a pluralistic perspective grounded in the conviction that 'the sciences are characteristically pluralistic, rather than unified, in language, subject matter, and method' (*ibid.*, p. 10), and that philosophy of science should be likewise pluralistic.

At the core of Suppes' probabilistic empiricism we find Bayesianism and subjectivism. References to de Finetti abound in Suppes' writings, but his position is in many ways more flexible than de Finetti's. A central feature of Suppes' perspective stems from the conviction that exact values should be substituted by probability intervals. In this connection, Suppes, partly in collaboration with Mario Zanotti, spelled out a cluster of results on upper and lower probabilities, giving rise to a most fruitful approach, which they regard

---

[42] As to Suppes' theory of probabilistic causality, see Suppes (1970) and (1984b); as to rationality see Suppes (1981).

as a natural extension of de Finetti's line of thought[43].

While agreeing with de Finetti on a number of points, including the tenet that probability and utility are notions to be dealt with separately, Suppes does not share de Finetti's uncompromising subjectivism. In particular, he objects to de Finetti's version of subjectivism that it cannot account for uncertain evidence, nor does it offer a way of evaluating different probability assessments due to experts. A major disagreement comes in connection with de Finetti's rejection of 'objective' probability, a notion that Suppes is instead willing to retain, and to combine with that of propensity[44].

A central feature of Suppes' pluralistic perspective lies in the fact that important concepts in science and philosophy are not given a univocal definition, being instead assigned a specific meaning depending on the context in which they occur. Probability is no exception to this tendency. While bringing to the fore the strict connection between probability and personal judgment and opinion, Suppes urges that in certain contexts probability should be attached an objective meaning. He makes the point that the question of the meaning of probability statements is not different from that of the meaning of statements about physical properties or magnitudes, like mass and weight. In all such cases, according to him 'there is not really an interesting and strong distinction between subjective and objective, or between belief and knowledge' (Suppes 1983, p. 399). The important thing appears instead to be completeness of information, and the relevant distinction to be made is that between complete knowledge in principle or in practice and incomplete knowledge with the possibility of learning more. When talking about the meaning of a probability statement, one has in the first place to ask whether it is based on complete information, in the sense that there is no additional information we can conditionalize on, that will bring about a change in the probability value. This is an important feature to be considered, especially when completeness of information comes from physical theory. In those contexts where this kind of completeness obtains, probability acquires an objective meaning.

To conclude this sketchy presentation of Suppes' probabilistic empiricism, it is worth mentioning that his position includes a constructivist view of scientific theories, according to which theories are sets of models intended to represent experimental evidence in the best possible way. Suppes' conception

---

[43] See the collection of their joint papers in Suppes and Zanotti (1996). The results on upper and lower probabilities by Suppes and Zanotti are summarized in Suppes (2002), Chapter 5.

[44] Suppes' views on propensity are surveyed in Section 5.3.

of theories is grounded on the conviction that a thorough analysis of scientific knowledge cannot ignore the statistical methodology for gathering experimental data and assessing hypotheses. The opening of philosophy of science to consideration of the statistical methods employed at the various levels of scientific investigation, including measurement, experimental design, estimation of parameters, tests of goodness of fit, identification of exogenous and endogenous variables, and the like, marks a decisive turn within philosophy of science, for a long time engaged only with the logical (syntactic and semantic) aspects of scientific knowledge. Suppes' philosophy of science supersedes the traditional distinction – which was due to Reichenbach and became a hallmark of logical empiricism – between a 'context of discovery', of exclusive concern of scientists, and a 'context of justification', of specific concern of philosophers of science. On the contrary, for Suppes philosophy of science should occupy itself with both contexts, a conviction which has come to be widely shared by present day philosophers of science. This implies a shift of emphasis from the logical and linguistic aspects of scientific knowledge – of preponderant interest for logical empiricists – to the statistical and probabilistic aspects of science, ranging from its experimental to its theoretical stage. Albeit in different ways, both Jeffrey and Suppes triggered a progressive opening of philosophy of science and epistemology to probability.

# *Closing remarks*

The preceding overview led us to examine a whole array of seemingly irreconcilable perspectives, rooted in some way or other in the duality of meaning that has imprinted modern probability since its beginning. The question whether any of the interpretations considered currently predominates over the others does not allow for a straightforward answer.

It seems undisputable that epistemic probability has a role to play in the realm of the social sciences where personal opinions and expectations enter directly into the information used to support forecasts, forge hypotheses and build models. As Donald Gillies observed, 'most of the principal advocates of the epistemological interpretation of probability (Keynes, Ramsey, de Finetti) were concerned with the application of probability in economics, and [...] most of the principal advocates of the objective interpretation of probability (von Mises, Fisher, Neyman, Popper) were concerned with the application of probability in the natural sciences (physics and biology)' (Gillies 2000, p. 187). While subjectivism is by and large accredited by economists, it may be claimed, with Gillies, that the case of economics can be extended to the social sciences in general, for 'financial markets are a pure case of the social' (*ibid.*, p. 194).

Conversely, the frequency interpretation, due to its empirical and objective character, has long been considered the natural candidate to account for the notion of probability occurring within the natural sciences. But while frequentism seems to match the uses of probability in areas like population genetics and statistical mechanics, quantum mechanics clashes with its fundamental assumption, namely the tenet that probability can only refer to populations, not to single events.

To solve this difficulty, in the Fifties Popper developed the propensity interpretation, according to which probability is a theoretical property of chance set-ups and expresses their tendency to exhibit frequencies that can be observed. In the debate that followed, propensity theory gained increasing popularity, but elicited various objections. In particular, it was pointed out that its single-case version is unsuitable to interpret inverse probabilities, and it faces a reference class problem analogous to that affecting the attempts which were made, for instance by Reichenbach, to formulate a version of frequentism applicable to the single case. One might get the impression that the problems of frequentism are displaced, rather than removed. All the more so, since according to both single-case and long-run versions propensity attributions represent conjectures to be tested against observed frequencies, making the applicability of the whole theory rest ultimately on frequencies.

What about the adoption of an epistemic approach to interpret probabilities in the natural sciences? Leaving aside the classical interpretation, whose tie with determinism makes it seem outdated, its direct descendant, namely the logical interpretation, should appeal to scientists, given its objectivistic flavour. As we saw, the viability of this alternative was maintained by the geophysicist Harold Jeffreys. In connection with quantum mechanical probabilities, Jeffreys held the view that they could be accounted for within the framework of epistemic probability, and their peculiar character identified with their being 'intrinsic' to the theory. Accordingly, their objectivity was made to descend from physical theories accepted by the scientific community. Albeit coming from a natural scientist, Jeffreys' methodology counts more followers among statisticians and economists than among physicists, and his philosophical ideas have received hardly any attention. As to other upholders of logicism, like Rudolf Carnap, his work is almost unknown outside the restricted circle of logicians and philosophers of science. Regarding logicism, as a general remark it can be observed that the pretence to defend a unique probability evaluation based on a given body of evidence has proved arduous.

If this is true, the natural move for those who embrace an epistemic conception of probability would be to turn to subjectivism, which is not committed to such a pretence. Unfortunately, Bruno de Finetti did not pay much attention to the problem of interpreting physical probabilities. He heralded an uncompromising form of subjectivism that does not contemplate a distinction between the assignments of probability regarding everyday life and those made in connection with science, although late in his life he seemed inclined to grant that scientific theories can give a sounder foundation to personal opinions. But de Finetti's view of scientific theories is crudely instrumental, and is not underpinned by a sophisticated philosophical background, comparable to that underlying Ramsey's pragmatic approach. Moreover, de Finetti opposed the notion of repeated experiments, and insisted that probability can only be referred to single events, or facts, as he used to say. While the possibility of ascribing probability to single events is undeniably an advantage of subjectivism over frequentism, de Finetti's refusal to consider repeated experiments resulted in a limitation of his approach. In addition, de Finetti embraced a phenomenalistic version of positivism, which he described as 'analogous to Mach's positivism, where by "positive fact" each of us means only his own subjective impressions' (de Finetti 1931b, English edition 1989, p. 171). Surely, this kind of philosophy did not help make de Finetti's subjectivism attractive to scientists[1].

Ramsey's subjectivism is more flexible, and accommodates within the subjectivist framework notions banned by de Finetti, such as those of chance and randomness. Furthermore, Ramsey develops a notion of 'probability in physics', grounded on the idea, emphasised by Jeffreys, that scientific theories have a direct bearing on probability assessments made by scientists. As we saw, a similar stance has been taken by Richard Jeffrey, and is progressively gaining the assent of other authors. In point of fact, the idea of attaching an epistemic interpretation to the uses of probability made by scientists is not at all implausible, especially since quantum mechanics raised problems in connection with frequentism, and even scientists of the prestige of Hermann Weyl and Johann von Neumann seem to have been open to this option[2]. Nevertheless, this kind of approach is far from popular among scientists like physicists and geneticists.

The controversy on the interpretation of physical probabilities is, then, all but settled. But the nature of statistical method is also the topic of lively debate. Although there is a broad consensus on the usefulness and even indispensability of statistical methodology in all branches of scientific research, the discussion on what is the 'right' approach in this connection is open and a thorough analysis of this issue would require another book. Suffice it to say

---

[1] See Galavotti (1995-96) for further comments on this point. See also Galavotti (2001b) for some remarks on the interpretation of physical probabilities.
[2] See the passages quoted in Suppes (2002), Chapter 5.

that the long-standing opposition between Bayesians and orthodox statisticians, who privilege the use of testing techniques, still divides statisticians, and even within these two leading schools there are deep divergencies. On the one hand, Bayesians are divided into subjectivists and objectivists, who take a different attitude as to the determination of priors; on the other, one can distinguish various currents among orthodox statisticians, depending on whether they privilege significance tests, Neyman-Pearson methodology, or other methods.

Far from being committed to a single method, researchers operating in various disciplines are inclined to adopt different methods, depending on circumstances. This eclectic and pragmatic attitude has been described by Suppes in his recent book *Representation and Invariance of Scientific Structures*, where he claims that 'by and large scientists in most disciplines remain indifferent to the conceptual foundations of probability and pragmatically apply statistical concepts without any foundational anxiety' (Suppes 2002, p. 263). Such attitude is counterbalanced by the pragmatic and pluralistic outlook of a number of epistemologists – including Suppes himself – who avoid taking a unilateral perspective on the foundations of probability and statistics, and more generally on the nature of scientific knowledge. Far from renouncing the search for a comprehensive theory meant to cover all uses of probability, the pragmatic approach should be seen as a constructive standpoint. Instead of settling on a given methodology or a univocal definition of probability, and then trying to force all uses of probability into a single scheme, the approach in question moves from the recognition of diversity to the detection of analogies. To this end a detailed analysis of the peculiarities characterizing different notions of probability and statistical methods is indispensable. This makes it all the more necessary to clarify the concept and uses of probability, in order to gain some understanding of scientific knowledge.

# References

Bacon, Francis (1620). *Novum Organum*. In *The 'Instauratio Magna'. Part II: 'Novum Organum' and Associated Texts*, edited and translated by Graham Rees and Maria Wakely. Oxford: Clarendon Press (The Oxford Francis Bacon, vol. 11), 2004.

Bacon, Francis (1857-74). *The Works of Francis Bacon*. 14 volumes, ed. by James Spedding, Robert Leslie Ellis and Douglas Denon Heath. London: Longmans.

Barnard, George (1958). 'Thomas Bayes – a biographical note. Together with a Reprinting of Bayes 1764'. *Biometrika* XLV, pp. 293–315. Reprinted in Pearson and Kendall, eds. (1970), pp. 131–53. Reprinted in Swinburne, ed. (2002), pp. 117–21.

Bayes, Thomas (1763). 'An Essay towards Solving a Problem in the Doctrine of Chances'. *Philosophical Transactions of the Royal Society* LIII, pp. 370–418. Reprinted in Pearson and Kendall, eds. (1970), pp. 131–53. Reprinted in Swinburne, ed. (2002), pp. 122–49.

Bernoulli, Daniel (1738). 'Specimen de theoriae novae de mensura sortis'. *Acta Academiae Scientiarum Imperialis Petropolitanae* V, pp. 175–92. English edition 'Exposition of a New Theory on the Measurement of Risk'. *Econometrica* XXII (1954), pp. 23–36.

Bernoulli, Daniel (1777). 'Diiudicatio maxime probabilis plurium observationum discrepantium atque verisimillima inductio inde formanda'. *Acta Academiae Scientiarum Imperialis Petropolitanae*, pp. 3–33. English edition 'The Most Probable Choice between Several Discrepant Observations and the Formation therefrom of the Most Likely Induction'. In Pearson and Kendall, eds. (1970), pp. 155–72 (includes a 'Foreword' by Maurice George Kendall and 'Observations on the Foregoing Dissertation of Bernoulli', by Leonhard Euler).

Bernoulli, Jakob (1713). *Ars conjectandi*. Basel. Reprinted in *Opera*, ed. by Gabriel Cramer. Geneva: Bousquet, 1742.

Bertrand, Joseph (1888). *Calcul des probabilités*. Paris: Gauthier-Villars. Second edition 1907; reprinted New York: Chelsea, 1972.

239

Bolzano, Bernard (1837). *Wissenschaftslehre*. Sulzbach: Seidel. English partial edition *Theory of Science*, ed. by Jan Berg. Dordrecht: Reidel, 1973.

Boole, George (1851). 'On the Theory of Probabilities, and in Particular on Mitchell's Problem of the Distribution of Fixed Stars'. *The Philosophical Magazine*, Series 4, I, pp. 521–30. Reprinted in Boole (1952), pp. 247–59.

Boole, George (1854a). *An Investigation of the Laws of Thought, on which are Founded the Mathematical Theories of Logic and Probabilities*. London: Walton and Maberly. Reprinted as *George Boole's Collected Works*, vol. 2. Chicago-New York: Open Court, 1916. Reprinted New York: Dover, 1951.

Boole, George (1854b). 'On a General Method in the Theory of Probabilities'. *The Philosophical Magazine*, Series 4, VIII, pp. 431–44. Reprinted in Boole (1952), pp. 291–307.

Boole, George (1952). *Studies in Logic and Probability*, ed. by Rush Rhees. London: Watts and Co.

Boole, George (1997). *Selected Manuscripts on Logic and its Philosophy*, ed. by Ivor Grattan-Guinness and Gérard Bornet. Berlin: Birkhäuser.

Borel, Émile (1924). 'À propos d'un traité des probabilités'. *Revue Philosophique* XCVIII, pp. 321–36. Reprinted in Borel (1972), vol. 4, pp. 2169–84. English edition 'Apropos of a Treatise on Probability'. In Kyburg and Smokler, eds. (1964), pp. 45–60, (not included in the second edition).

Borel, Émile (1972). *Oeuvres de Émile Borel*. 4 volumes. Paris: Éditions du CNRS.

Box, Joan Fisher (1978). *R.A. Fisher: The Life of a Scientist*. New York: Wiley.

Braithwaite, Richard Bevan (1946). 'John Maynard Keynes, First Baron Keynes of Tilton'. *Mind* LV, pp. 283–4.

Braithwaite, Richard Bevan (1975). 'Keynes as a Philosopher'. In Milo Keynes, ed. (1975), pp. 237–46.

Broad, Charlie Dunbar (1922). 'Critical Notices: A Treatise on Probability by J.M. Keynes'. *Mind* XXXI, pp. 72–85.

Broad, Charlie Dunbar (1924). 'Mr. Johnson on the Logical Foundations of Science'. *Mind* XXXIII, pp. 242–69; 367–84.

Broad, Charlie Dunbar (1931). 'William Ernest Johnson'. *Proceedings of the British Academy* XVII, pp. 491–514.

Butts, Robert (1973). 'Whewell's Logic of Induction'. In Giere and Westfall, eds. (1973), pp. 53–85.

Byrne, Edmund (1968). *Probability and Opinion. Study in the Medieval Presuppositions of Post-Medieval Theories of Probability*. The Hague: Martinus Nijhoff.

Cameron, Laura and Forrester, John (2000). 'Tansley's Psychoanalytic Network: An Episode out of the Early History of Psychoanalysis in England'. *Psychoanalysis and History* II, pp. 189–256.

Campbell, N.R. (1920). *Physics, the Elements*. Cambridge: Cambridge University Press. Reprinted as *Foundations of Science*, New York: Dover, 1957.

Carabelli, Anna (1988). *On Keynes's Method*. London: Macmillan.

Carnap, Rudolf (1945a). 'The Two Concepts of Probability'. *Philosophy and Phenomenological Research* V, pp. 513–32. Reprinted in Feigl and Sellars, eds.

(1949), pp. 330–48.

Carnap, Rudolf (1945b). 'On Inductive Logic'. *Philosophy of Science* XII, pp. 72–97. Reprinted in Luckenbach, ed. (1972), pp. 51–79.

Carnap, Rudolf (1946). 'Remarks on Induction and Truth'. *Philosophy and Phenomenological Research* VI, pp. 590–602.

Carnap, Rudolf (1947a). 'Probability as a Guide in Life'. *The Journal of Philosophy* XLIV, pp.141–8.

Carnap, Rudolf (1947b). 'On the Application of Inductive Logic'. *Philosophy and Phenomenological Research* VIII, pp. 133–48.

Carnap, Rudolf (1949). 'Truth and Confirmation'. In Feigl and Sellars, eds. (1949), pp. 119–27.

Carnap, Rudolf (1950). *Logical Foundations of Probability*. Chicago: Chicago University Press. Second edition with modifications 1962, reprinted 1967.

Carnap, Rudolf (1952). *The Continuum of Inductive Methods*. Chicago: Chicago University Press.

Carnap, Rudolf (1953). 'Inductive Logic and Science'. *Proceedings of the American Academy of Arts and Sciences* LIII, pp. 189–97.

Carnap, Rudolf (1962). 'The Aim of Inductive Logic'. In *Logic, Methodology and Philosophy of Science*, ed. by Ernest Nagel, Patrick Suppes and Alfred Tarski. Stanford: Stanford University Press, pp. 303–18. Reprinted in Luckenbach, ed. (1972), pp. 104–20.

Carnap, Rudolf (1963a). 'Remarks on Probability'. *Philosophical Studies* XIV, pp. 65-75.

Carnap, Rudolf (1963b). 'Intellectual Autobiography'. In Schilpp, ed. (1963), pp. 3–84.

Carnap, Rudolf (1963c). 'Replies and Systematic Expositions'. In Schilpp, ed. (1963), pp. 859–1013.

Carnap, Rudolf (1968). 'Inductive Logic and Inductive Intuition'. In Lakatos, ed. (1968), pp. 258–67.

Carnap, Rudolf (1971). 'Inductive Logic and Rational Decisions'. In Carnap and Jeffrey, eds. (1971), pp. 7–31.

Carnap, Rudolf (1975). 'Notes on Probability and Induction'. In Hintikka, ed. (1975), pp. 293–324.

Carnap, Rudolf (1980). 'A Basic System of Inductive Logic, Part II'. In Jeffrey, ed. (1980), pp. 7–156.

Carnap, Rudolf; Hahn, Hans and Neurath, Otto (1929). *Wissenschaftliche Weltauffassung. Der Wiener Kreis*. Vienna: A. Wolf. English edition 'The Scientific Conception of the World'. In Otto Neurath, *Empiricism and Sociology*, ed. by Maria Neurath and Robert S. Cohen. Dordrecht: Kluwer, 1973, pp. 299–318.

Carnap, Rudolf and Jeffrey, Richard C., eds. (1971). *Studies in Inductive Logic and Probability*, volume 1. Berkeley-Los Angeles-London: University of California Press.

Church, Alonzo (1940). 'The Concept of a Random Sequence'. *Bulletin of the American Mathematical Society* XLVI, pp. 130–5.

Clark, Peter (1995). 'Popper on Determinism'. In *Karl Popper: Philosophy and Problems*, ed. by Anthony O'Hear. Cambridge: Cambridge University Press, pp. 149–62.

Cook, Alan (1990). 'Sir Harold Jeffreys'. *Biographical Memoirs of Fellows of the Royal Society* XXXVI, pp. 303–33.

Costantini, Domenico and Galavotti, Maria Carla (1987). 'Johnson e l'interpretazione degli enunciati probabilistici'. In *L'epistemologia di Cambridge 1850-1950*, ed. by Raffaella Simili. Bologna: Il Mulino, pp. 245–62.

Costantini, Domenico and Galavotti, Maria Carla, eds. (1997). *Probability, Dynamics and Causality*. Dordrecht-Boston: Kluwer.

Cournot, Antoine Augustin (1843). *Exposition de la théorie des chances et des probabilités*. Paris: Hachette. Also in *Oeuvres complètes*. Paris: Vrin, 1984.

Dale Andrew (1991). *A History of Inverse Probability. From Thomas Bayes to Karl Pearson*. New York: Springer.

Daston, Lorraine (1988). *Classical Probability in the Enlightenment*. Princeton: Princeton University Press.

David, Florence Nightingale (1955). 'Dicing and Gaming (A Note on the History of Probability)'. *Biometrika* XLII, pp. 1–15. Reprinted in Pearson and Kendall, eds. (1970), pp. 1–17.

David, Florence Nightingale (1962). *Games, God and Gambling*. London: Griffin.

Dawid, Philip (1994). 'Foundations of Probability'. In *Companion Encyclopedia of the History and Philosophy of the Mathematical Sciences*, ed. by Ivor Grattan-Guinness. London-New York: Routledge. Vol. 2, pp. 1399–406.

de Finetti, Bruno (1930). 'Fondamenti logici del ragionamento probabilistico'. *Bollettino dell'Unione Matematica Italiana* V, pp. 1–3. Reprinted in de Finetti (1981), pp. 261–3.

de Finetti, Bruno (1931a). 'Sul significato soggettivo della probabilità'. *Fundamenta mathematicae* XVII, pp. 298–329. English edition 'On the Subjective Meaning of Probability'. In de Finetti (1992), pp. 291–321.

de Finetti, Bruno (1931b). 'Probabilismo'. *Logos*, pp. 163–219. Reprinted in de Finetti (1989), pp. 3–70. English edition 'Probabilism'. In *Erkenntnis* XXXI (1989), pp. 169–223.

de Finetti, Bruno (1931c). 'Le leggi differenziali e la rinuncia al determinismo'. *Rendiconti del Seminario Matematico della R. Università di Roma*, serie 2, VII, pp. 63–74. English edition 'Differential Laws and the Renunciation of Determinism'. In de Finetti (1992), pp. 323–34.

de Finetti, Bruno (1933). 'Sul concetto di probabilità'. *Rivista italiana di statistica, economia e finanza* V, pp. 723–47. English edition 'On the Probability Concept'. In de Finetti (1992), pp. 335–52.

de Finetti, Bruno (1936). 'Statistica e probabilità nella concezione di R. von Mises'. *Supplemento statistico ai Nuovi problemi di Politica, Storia, ed Economia* II, pp. 9–19. English edition 'Statistics and Probability in R. von Mises' Conception'. In de Finetti (1992), pp. 353–64.

de Finetti, Bruno (1937). 'La prévision: ses lois logiques, ses sources subjectives'. *Annales de l'Institut Henri Poincaré* VII, pp.1–68. English edition 'Foresight: its Logical Laws, its Subjective Sources'. In Kyburg and Smokler, eds. (1964), pp. 95–158. Also in the second edition (1980), pp. 53–118.

de Finetti, Bruno (1938a). 'Probabilisti di Cambridge'. *Supplemento statistico ai Nuovi problemi di Politica, Storia, ed Economia* IV, pp. 21–37. English edition 'Cambridge Probability Theorists'. *Rivista di matematica per le scienze economiche e sociali* VIII (1985), pp. 79–91.

de Finetti, Bruno (1938b). 'Resoconto critico del colloquio di Ginevra intorno alla teoria della probabilità'. *Giornale dell'Istituto Italiano degli Attuari* IX, pp. 3–42.

de Finetti, Bruno (1939). 'Punti di vista: Émile Borel'. *Supplemento statistico ai Nuovi problemi di Politica, Storia, ed Economia* V, pp. 61–71.

de Finetti, Bruno (1941). 'Punti di vista: Hans Reichenbach'. *Statistica* I, pp. 125–33.

de Finetti, Bruno (1951a). 'Recent Suggestions for the Reconciliation of Theories of Probability'. In *Proceedings of the Second Berkeley Symposium on Mathematical Statistics and Probability*, ed. by Jerzy Neyman. Berkeley: University of California Press, pp. 217–25.

de Finetti, Bruno (1951b). 'L'opera di Abraham Wald e l'assestamento concettuale della statistica matematica moderna'. *Statistica* XI, pp. 185–92.

de Finetti Bruno (1957). 'L'informazione, il ragionamento, l'inconscio nei rapporti con la previsione'. *L'industria* II, pp. 3–27.

de Finetti, Bruno (1962). 'Does it Make Sense to Speak of "Good Probability Appraisers"?'. In *The Scientist Speculates. An Anthology of Partly-Baked Ideas*, ed. by Irving John Good, Alan James Mayne and John Maynard Smith. New York: Basic Books, pp. 357–64.

de Finetti, Bruno (1963), 'La decisione nell'incertezza'. *Scientia* XCVIII, pp. 61–8.

de Finetti, Bruno (1967). 'L'adozione della concezione soggettivistica come condizione necessaria e sufficiente per dissipare secolari pseudoproblemi'. In *I fondamenti del calcolo delle probabilità*, ed. by Dario Fürst and Giuseppe Parenti. Florence: Scuola di Statistica dell'Università, pp. 57–92. Replies by de Finetti in the 'Appendice', pp. 93–245.

de Finetti, Bruno (1968). 'Probability: the Subjectivistic Approach'. In *La philosophie contemporaine*, ed. by Raymond Klibansky. Florence: La Nuova Italia, pp. 45–53.

de Finetti, Bruno (1969). 'Initial Probabilities: a Prerequisite for any Valid Induction'. *Synthèse* XX, pp. 2–16.

de Finetti, Bruno (1970). *Teoria delle probabilità*, Torino: Einaudi. English edition *Theory of Probability*. New York: Wiley, 1975.

de Finetti, Bruno (1972a). 'Subjective or Objective Probability: is the Dispute Undecidable?'. *Symposia Mathematica* IX, pp. 21–36.

de Finetti, Bruno (1972b). *Probability, Induction and Statistics*. New York: Wiley.

de Finetti, Bruno (1973). 'Bayesianism: Its Unifying Role for Both the Foundations and the Applications of Statistics'. *Bulletin of the International Statistical Institute, Proceedings of the 39th Session*, pp. 349–68.

de Finetti, Bruno (1974). 'The Value of Studying Subjective Evaluations of Probability', in *The Concept of Probability in Psychological Experiments*, ed. by Carl-Axel Staël von Holstein. Dordrecht-Boston: Reidel, pp. 1–14.

de Finetti, Bruno (1976). 'Probability: Beware of Falsifications!'. *Scientia* LXX, pp. 283–303. Reprinted in Kyburg and Smokler, eds. (1964), second edition (1980), pp. 194–224 (not in the first edition).

de Finetti, Bruno (1980). 'Probabilità'. In *Enciclopedia Einaudi*. Torino: Einaudi. Vol. 10, pp. 1146-87.

de Finetti, Bruno (1981). *Scritti (1926-1930)*. Padua: CEDAM.

de Finetti, Bruno (1982). 'Probability and my Life'. In *The Making of Statisticians*, ed. by Joseph Gani. New York: Springer, pp. 4–12.

de Finetti, Bruno (1989). *La logica dell'incerto*, ed. by Marco Mondadori. Milan: Il Saggiatore.

de Finetti, Bruno (1991). *Scritti (1931-1936)*. Bologna: Pitagora.

de Finetti, Bruno (1992). *Probabilità e induzione (Induction and Probability)*, ed. by Paola Monari and Daniela Cocchi. Bologna: CLUEB (a collection of de Finetti's papers both in Italian and in English).

de Finetti, Bruno (1995). *Filosofia della probabilità*, ed. by Alberto Mura. Milan: Il Saggiatore.

de Finetti, Bruno and Savage, Leonard Jimmie (1962). 'Sul modo di scegliere le probabilità iniziali'. *Biblioteca del Metron. Serie C: Note e commenti*. Rome: Istituto di Statistica dell'Università, pp. 82–154.

de Finetti, Fulvia (2000). 'Alcune lettere giovanili di B. de Finetti alla madre'. *Nuncius* XV, pp. 721–40.

De Morgan, Augustus (1837). 'Theory of Probabilities'. In *Encyclopaedia Metropolitana*.

De Morgan, Augustus (1838). *An Essay on Probabilities, and on their Applications to Life, Contingencies and Insurance Offices*. London: Longman.

De Morgan, Augustus (1847). *Formal Logic: or, The Calculus of Inference, Necessary and Probable*. London: Taylor and Walton. Reprinted London: Open Court, 1926.

De Morgan, Sophia Elizabeth (1882). *Memoir of Augustus De Morgan*. London: Longman.

d'Holbach, Paul Henri Thiry (1770). *Système de la nature, ou des lois du monde physique et du monde moral*. London-Amsterdam: Rey. English edition *The System of Nature*. New York: Franklin, 1970 (from the translation of H.D. Robinson, London, 1868).

Di Maio, Maria Concetta (1994). 'Review of F.P. Ramsey, *Notes on Philosophy, Probability and Mathematics*'. *Philosophy of Science* LXI, pp. 487–9.

Dokic, Jérôme and Engel, Pascal (2001). *Vérité et Succès*. Paris: Presses Universitaires de France. English edition *Frank Ramsey. Truth and Success*. London-New York: Routledge, 2003.

Donkin, William (1851). 'On Certain Questions Relating to the Theory of Probabilities'. *The Philosophical Magazine*, Series IV, I, pp. 353–68, 458–66; II, pp. 55–60.

Dummett, Michael (1993). *Origins of Analytical Philosophy*. London: Duckworth.

Dunnington, Guy Waldo (1955). *Carl Friedrich Gauss, Titan of Science; a Study of his Life and Work*. New York: Exposition Press.

Earman, John (1986). *A Primer on Determinism*. Dordrecht: Kluwer.

Earman, John (1992). *Bayes or Bust? A Critical Examination of Bayesian Confirmation Theory*. Cambridge, Mass.: The MIT Press.

Earman, John (2000). *Hume's Abject Failure*. Oxford: Oxford University Press.

Earman, John and Wesley C. Salmon (1992). 'The Confirmation of Scientific Hypotheses'. In Merrilee H. Salmon, John Earman, Clark Glymour, James G. Lennox, Peter Machamer, J.E. McGuire, John D. Norton, Wesley C. Salmon and Kenneth Schaffner, *Introduction to the Philosophy of Science*. Englewood Cliffs, New Jersey: Prentice Hall, pp. 42–103.

Edwards, Anthony William Fairbank (1972). *Likelihood*. Cambridge: Cambridge University Press.

Ellis, Robert Leslie (1849). 'On the Foundations of the Theory of Probability'. *Transactions of the Cambridge Philosophical Society* VIII, pp. 1–6. Reprinted in Ellis (1863), pp. 1–11.

Ellis, Robert Leslie (1856). 'Remarks on the Fundamental Principle of the Theory of Probabilities'. *Transactions of the Cambridge Philosophical Society* IX, pp. 605–7. Reprinted in Ellis (1863), pp. 49–52.

Ellis, Robert Leslie (1863). *The Mathematical and Other Writings*, ed. by William Walton. Cambridge: Deighton, Bell and Co.

Feigl, Herbert (1950). 'De Principiis non Disputandum ...? On the Meaning and the Limits of Justification'. In *Philosophical Analysis*, ed. by Max Black. Ithaca, N.Y.: Cornell University Press, pp. 119–56. Reprinted in *Inquiries and Provocations: Selected Writings 1929-1974*, ed. by Robert S. Cohen. Dordrecht: Reidel, 1980, pp. 237–68.

Feigl, Herbert and Sellars, Wilfrid, eds. (1949). *Readings in Philosophical Analysis*. New York: Appleton-Century-Crofts.

Feller, William (1950), (1966). *An Introduction to Probability Theory and its Applications*. New York: Wiley. Vol. 1, 1950; vol. 2, 1966.

Fermat, Pierre (1679). *Varia opera mathematica*. Toulouse: Peck.

Festa, Roberto (1993). *Optimum Inductive Methods*. Dordrecht-Boston: Kluwer.

Festa, Roberto (1999). 'Bayesian Confirmation'. In *Experience, Reality, and Scientific Explanation*, ed. by Maria Carla Galavotti and Alessandro Pagnini. Dordrecht-Boston: Kluwer, pp. 55–88.

Fetzer, James (1974a). 'Statistical Probabilities: Single Case Propensities versus Long-Run Frequencies'. In *Developments in the Methodology of Social Sciences*, ed. by Werner Leinfellner and Eckehart Köhler. Dordrecht-Boston: Reidel, pp. 387–97.

Fetzer, James (1974b). 'A Single Case Propensity Theory of Explanation'. *Synthèse* XXVIII, pp. 171–98.

Fisher, Ronald Aylmer (1925). *Statistical Methods for Research Workers*. Edinburgh: Oliver and Boyd.

Fisher, Ronald Aylmer (1930). *The Genetical Theory of Natural Selection*. Oxford: Clarendon Press.

Fisher, Ronald Aylmer (1935). *The Design of Experiments*. Edinburgh: Oliver and Boyd.

Francis, Henry Thomas (1923). 'In Memoriam: John Venn'. *The Caian* XXXI. Reprinted Cambridge University Press.

Franklin, James William (2001). *The Science of Conjecture: Evidence and Probability before Pascal*. Baltimore-London: The Johns Hopkins University.

Fuller, Jean Overton (1981). *Francis Bacon: A Biography*. London: East-West

Publications.

Galavotti, Maria Carla (1989). 'Anti-realism in the Philosophy of Probability: Bruno de Finetti's Subjectivism'. *Erkenntnis* XXXI, pp. 239–61.

Galavotti, Maria Carla (1991). 'The Notion of Subjective Probability in the Work of Ramsey and de Finetti'. *Theoria* LVII, pp. 239–59.

Galavotti, Maria Carla (1994). 'Some Observations on Patrick Suppes' Philosophy of Science'. In Paul Humphreys, ed. *Patrick Suppes: Scientific Philosopher.* Dordrecht-Boston: Kluwer. Vol. 3, pp. 245–70.

Galavotti, Maria Carla (1995). 'F.P. Ramsey and the Notion of "Chance"'. In *The British Tradition in the 20$^{th}$ Century Philosophy. Proceedings of the 17$^{th}$ International Wittgenstein Symposium*, ed. by Jaakko Hintikka and Klaus Puhl. Vienna: Holder-Pichler-Tempsky, pp. 330–40.

Galavotti, Maria Carla (1995-96). 'Operationism, Probability and Quantum Mechanics'. *Foundations of Science* I, pp. 99–118.

Galavotti, Maria Carla (1996). 'Probabilism and Beyond'. *Erkenntnis* XLV, pp. 253–65. Reprinted in Costantini and Galavotti, eds. (1997), pp. 113–26.

Galavotti, Maria Carla (1999). 'Some Remarks on Objective Chance (F.P. Ramsey, K.R. Popper and N.R. Campbell)'. In *Language, Quantum, Music*, ed. by Maria Luisa Dalla Chiara, Roberto Giuntini and Federico Laudisa. Dordrecht-Boston: Kluwer, pp. 73–82.

Galavotti, Maria Carla (2001a). 'Subjectivism, Objectivism and Objectivity in Bruno de Finetti's Bayesianism'. In *Foundations of Bayesianism*, ed. by David Corfield and Jon Williamson. Dordrecht-Boston: Kluwer, pp. 161–74.

Galavotti, Maria Carla (2001b). 'What Interpretation for Probability in Physics?'. In *Chance in Physics*, ed. by Jean Bricmont, Detlev Dürr, Maria Carla Galavotti, Giancarlo Ghirardi, Francesco Petruccione, Nino Zanghì. Berlin: Springer, pp. 265–70.

Galavotti, Maria Carla (2003). 'Harold Jeffreys' Probabilistic Epistemology: Between Logicism and Subjectivism'. *British Journal for the Philosophy of Science* LIV, pp. 43–57.

Galton, Francis (1869). *Hereditary Genius: An Inquiry into its Laws and Consequences.* London: Macmillan.

Galton, Francis (1883). *Inquiries into Human Faculty and its Development.* London: Macmillan.

Galton, Francis (1907). *Probability, the Foundation of Eugenics.* Oxford: Clarendon Press.

Garber, Daniel and Zabell, Sandy (1979). 'On the Emergence of Probability'. *Archive for History of Exact Sciences* XXI, pp. 33–53.

Giere, Ronald (1973). 'Objective Single-Case Probabilities and the Foundations of Statistics'. In *Logic, Methodology and Philosophy of Science IV*, ed. by Patrick Suppes, Leon Henkin, Athanase Joja and Gr. C. Moisil. Amsterdam-London: North-Holland, pp. 467–83.

Giere, Ronald (1976). 'A Laplacean Formal Semantics for Single-Case Propensities'. *Journal of Philosophical Logic* V, pp. 321–53.

Giere, Ronald N. and Westfall, Richard S., eds. (1973). *Foundations of Scientific*

*Method: the Nineteenth Century*. Bloomington-London: Indiana University Press.

Gigerenzer, Gerd; Swijtink, Zeno; Porter, Theodore; Daston, Lorraine; Beatty, John and Krüger, Lorenz (1989). *The Empire of Chance*. Cambridge: Cambridge University Press.

Gillham, Nicholas Wright (2001). *A Life of Sir Francis Galton. From African Explorations to the Birth of Eugenics*. Oxford: Oxford University Press.

Gillies, Donald (1988). 'Keynes as a Methodologist'. *British Journal for the Philosophy of Science* XXXIX, pp. 117–29.

Gillies, Donald (1998). 'Confirmation Theory'. In *Handbook of Defeasible Reasoning and Uncertainty Management Systems*, ed. by Dov Gabbay and Philippe Smets. Dordrecht-Boston: Kluwer. Vol. 1, pp. 135–67.

Gillies, Donald (2000). *Philosophical Theories of Probability*. London-New York: Routledge.

Gillispie, Charles Coulston, ed. (1972). *Dictionary of Scientific Biography*. New York: Charles Scribner's Sons.

Gillispie, Charles Coulston (1997). *Pierre-Simon Laplace, 1749-1827: A Life in Exact Science*. Chichester: Princeton University Press. (With the collaboration of Robert Fox and Ivor Grattan-Guinness).

Gnedenko, Boris Vladimirovich and Khinchin, Alexandr Yakovlevich (1962). *An Elementary Introduction to the Theory of Probability*. New York: Dover. Translated from the fifth Russian edition (Moscow, 1960) by Leo F. Boron and Sidney F. Mack.

Good, Irving John (1965). *The Estimation of Probabilities. An Essay on Modern Bayesian Methods*. Cambridge, Mass.: MIT Press.

Good, Irving John (1967). 'On the Principle of Total Evidence'. *British Journal for the Philosophy of Science* XVII, pp. 319–21.

Good, Irving John (1983). *Good Thinking. The Foundations of Probability and its Applications*. Minneapolis: University of Minnesota Press.

Good, Irving John (1988). 'Scientific Method and Statistics'. In *Encyclopedia of Statistical Sciences*, ed. by Samuel Kotz and Norman Johnson. New York: Wiley. Vol. 8, pp. 291–304.

Granger, Gilles-Gaston (1956). *La mathématique sociale du Marquis de Condorcet*. Paris, 1956. Second edition Paris, Editions Odile Jacob, 1989.

Graunt, John (1662). *Natural and Political Observations Made upon the Bills of Mortality*. London: Martin.

Hacking, Ian (1965). *Logic of Statistical Inference*. Cambridge: Cambridge University Press.

Hacking, Ian (1971a). 'The Leibniz-Carnap Program for Inductive Logic'. *The Journal of Philosophy* LXVIII, pp. 597–610.

Hacking, Ian (1971b). 'Equipossibility Theories of Probability'. *British Journal for the Philosophy of Science* XXII, pp. 339–55.

Hacking, Ian (1975). *The Emergence of Probability*. Cambridge: Cambridge University Press.

Hacking, Ian (1980). 'Grounding Probabilities from Below'. In *PSA 1980*, vol. 1, ed. by Peter Asquith and Ronald Giere. East Lansing: Philosophy of Science

Association, pp. 110–6.

Hacking, Ian (1990). *The Taming of Chance*. Cambridge: Cambridge University Press.

Hacohen, Malachi Haim (2000). *Karl Popper. The Formative Years 1902-1945*. Cambridge: Cambridge University Press.

Hahn, Roger (1967). *Laplace as a Newtonian Scientist*. William Andrews Clark Memorial Library, UCLA.

Hailperin, Theodor (1976). *Boole's Logic and Probability*. Amsterdam: North Holland.

Hammond, Nicholas, ed. (2003). *The Cambridge Companion to Pascal*. Cambridge: Cambridge University Press.

Hampshire, Stuart (1960). 'Friedrich Waismann'. *Proceedings of the British Academy* XLVI, pp. 310–7.

Harrod, Roy Forbes (1951). *The Life of John Maynard Keynes*. London: Macmillan.

Heidelberger, Michael (2001). 'Origins of the Logical Theory of Probability: von Kries, Wittgenstein, Waismann'. *International Studies in the Philosophy of Science* XV, pp. 177–88.

Heisenberg, Werner (1958). *The Physicist's Conception of Nature*. London: Hutchinson.

Heisenberg, Werner (1959). *Physics and Philosophy*. London: Allen and Unwin. Penguin edition 1990.

Hempel, Carl Gustav (1935). 'Über der Gehalt von Wahrscheinlichkeitsaussagen'. *Erkenntnis* V, pp. 228-60. English edition 'On the Content of Probability Statements'. In Hempel (2000), pp. 89–123.

Hempel, Carl Gustav (2000). *Selected Philosophical Essays*, ed. by Richard C. Jeffrey. Cambridge: Cambridge University Press.

Herschel, John (1830). *Preliminary Discourse on the Study of Natural Philosophy*. London: Longman. Reprinted London: Thoemmes, 1996.

Hintikka, Jaakko (1966). 'A Two-dimensional Continuum of Inductive Methods'. In *Aspects of Inductive Logic*, ed. by Jaakko Hintikka and Patrick Suppes. Amsterdam: North-Holland, pp. 113–32.

Hintikka, Jaakko, ed. (1975). *Rudolf Carnap, Logical Empiricist*. Dordrect: Reidel.

Holland, J.D. (1962). 'The Reverend Thomas Bayes, F.R.S. (1702-61)'. *Journal of the Royal Statistical Society, Series A* CXXV, pp. 451–61.

Howie, David (2002). *Interpreting Probability*. Cambridge: Cambridge University Press.

Howson, Colin (1988). 'On the Consistency of Jeffreys's Simplicity Postulate, and its Role in Bayesian Inference'. *The Philosophical Quarterly* XXXVIII, pp. 68–83.

Howson, Colin (1996). 'Bayesian Rules of Updating'. *Erkenntnis* XLV, pp. 195–208. Reprinted in Costantini and Galavotti, eds. (1997), pp. 55–68.

Howson, Colin and Urbach, Peter (1989). *Scientific Reasoning. The Bayesian Approach*. La Salle, Illinois: Open Court. Second modified edition 1993.

Hull, David (1973). 'Charles Darwin and the Nineteenth-Century Philosophies of Science'. In Giere and Westfall, eds. (1973), pp. 115–32.

Hume, David (1739). *A Treatise on Human Nature*, London. 1962 edition by D.G.C. Macnabb, Oxford: Collins.

Hume, David (1748). *An Enquiry Concerning Human Understanding*. London. 1999 edition by Tom L. Beauchamp. Oxford: Oxford University Press.

Humphreys, Paul (1985). 'Why Propensities Cannot be Probabilities'. *Philosophical Review* XCIV, pp. 557–70.

Humphreys, Paul, ed. (1994). *Patrick Suppes: Scientific Philosopher*. 3 volumes. Dordrecht-Boston: Kluwer.

Huygens, Christiaan (1888-1950). *Oeuvres complètes de Christiaan Huygens*. 22 volumes. The Hague: Dutch Academy of Sciences.

Ismael, Jenann (1996). 'What Chances Could not Be'. *British Journal for the Philosophy of Science* XLVII, pp. 79–91.

Jaynes, Edwin T. (1983). *Papers on Probability, Statistics and Statistical Physics*, ed. by Roger Rosenkrantz. Dordrecht: Reidel.

Jaynes, Edwin T. (2003). *Probability Theory: The Logic of Science*, ed. by G. Larry Bretthorst. Cambridge: Cambridge University Press.

Jeffrey, Richard (1965). *The Logic of Decision*. Chicago: The University of Chicago Press. Second edition Chicago: The University of Chicago Press, 1983.

Jeffrey, Richard C. (1975). 'Carnap's Empiricism'. In *Induction, Probability and Confirmation*, ed. by Grover Maxwell and Robert Milford Anderson. Minneapolis: University of Minnesota Press, pp. 37–49.

Jeffrey, Richard C., ed. (1980). *Studies in Inductive Logic and Probability*, volume 2. Berkeley-Los Angeles-London: University of California Press.

Jeffrey, Richard C. (1991). 'After Carnap'. *Erkenntnis* XXXV, pp. 255–62.

Jeffrey, Richard C. (1992a). *Probability and the Art of Judgment*. Cambridge: Cambridge University Press.

Jeffrey, Richard C. (1992b). 'Radical Probabilism (Prospectus for a User's Manual)'. In *Rationality in Epistemology*, ed. by Enrique Villanueva. Atascadero, Cal.: Ridgeview, pp. 193–204.

Jeffrey, Richard C. (1993). 'De Finetti's Radical Probabilism'. In de Finetti (1992), pp. 263–75.

Jeffrey, Richard C. (1996). 'Unknown Probabilities'. *Erkenntnis* XLV, pp. 327–35. Reprinted in Costantini and Galavotti, eds. (1997), pp. 187–95.

Jeffrey, Richard C. (2004). *Subjective Probability: The Real Thing*. Cambridge: Cambridge University Press.

Jeffreys, Harold (1922). 'Review of J.M. Keynes, *A Treatise on Probability*'. *Nature* CIX, pp. 132–3. Also in *Collected Papers* VI, pp. 253–6.

Jeffreys, Harold (1931). *Scientific Inference*. Cambridge: Cambridge University Press. Reprinted with 'Addenda' 1937, 2nd modified edition 1957, 1973.

Jeffreys, Harold (1933). 'Probability, Statistics and the Theory of Errors'. *Proceedings of the Royal Society*, Series A, CXL, pp. 523–35.

Jeffreys, Harold (1934). 'Probability and Scientific Method'. *Proceedings of the Royal Society*, Series A, CXLVI, pp. 9–16.

Jeffreys, Harold (1936a). 'The Problem of Inference'. *Mind* XLV, pp. 324–33.

Jeffreys, Harold (1936b). 'On Some Criticisms of the Theory of Probability'. *Philosophical Magazine* XXII, pp. 337–59.

Jeffreys, Harold (1937). 'Scientific Method, Causality, and Reality'. *Proceedings of the Aristotelian Society*, New Series, XXXVII, pp. 61–70.

Jeffreys, Harold (1939). *Theory of Probability*. Oxford, Clarendon Press, 2nd modified edition 1948, 1961, 1983.

Jeffreys, Harold (1955). 'The Present Position in Probability Theory'. *The British Journal for the Philosophy of Science* V, pp. 275–89. Also in *Collected Papers* VI, pp. 421–35.

Jeffreys, Harold and Swirles, Bertha (1971-1977). *Collected Papers of Sir Harold Jeffreys on Geophysics and Other Sciences*. 6 volumes. London-Paris-New York: Gordon and Breach Science Publishers.

Jevons, William Stanley (1873). *The Principles of Science*. London: Macmillan. Second enlarged edition 1877. Reprinted New York 1958.

Johnson, William Ernest (1921, 1922, 1924). *Logic.* Cambridge: Cambridge University Press. *Part I*, 1921; *Part II*, 1922; *Part III*, 1924. Reprinted New York: Dover, 1964.

Johnson, William Ernest (1932). 'Probability: The Relations of Proposal to Supposal; Probability: Axioms; Probability: The Deductive and the Inductive Problems'. *Mind* XLI, pp. 1–16, 281–96, 409–23.

Jorland, Gérard (1987). 'The Saint Petersburg Paradox 1713-1937'. In Krüger, Gigerenzer and Morgan, eds. (1987), vol. 1, pp. 157–190.

Kamlah, Andreas (1983). 'Probability as a Quasi-Theoretical Concept: J. von Kries' Sophisticated Account after a Century'. *Erkenntnis* XIX, pp. 239–51.

Kamlah, Andreas (1987). 'The Decline of the Laplacian Theory of Probability: A Study of Stumpf, von Kries, and Meinong'. In Krüger, Gigerenzer and Morgan, eds. (1987), vol. 1, pp. 91–116.

Kantola, Ilkka (1994). *Probability and Moral Uncertainty in late Medieval and Early Modern Times*. Helsinki: Luther-Agricola-Society.

Kendall, Maurice George (1963). 'Ronald Aylmer Fisher 1890-1962'. *Biometrika* L, pp. 1–15. Reprinted in Pearson and Kendall, eds. 1970, pp. 439–53.

Kendall, Maurice George (1941). 'A Theory of Randomness'. *Biometrika* XXXII, pp. 1–15.

Kenny, Anthony (1973). *Wittgenstein*. London: Pelican Books.

Keynes, John Maynard (1921). *A Treatise on Probability*. London: Macmillan. Reprinted in Keynes (1972), vol. 8.

Keynes, John Maynard (1930). 'Frank Plumpton Ramsey'. *The Economic Journal.* Reprinted in Keynes (1933) and (1972), pp. 335–46.

Keynes, John Maynard (1931). 'W.E. Johnson'. *The Times*, 15 January 1931. Reprinted in Keynes (1933) and (1972), pp. 349-50.

Keynes, John Maynard (1933). *Essays in Biography*. London: Macmillan. Second modified edition 1951. Third modified edition in Keynes (1972), vol. 10.

Keynes, John Maynard (1936). 'William Stanley Jevons'. *Journal of the Royal Statistical Society* Part III. Reprinted in the second edition of Keynes (1933) and in (1972), pp. 109–60.

Keynes, John Maynard (1972). *The Collected Writings of John Maynard Keynes*. Cambridge: Macmillan.

Keynes, Milo, ed. (1975). *Essays on John Maynard Keynes*. Cambridge: Cambridge University Press.

Kneale, William (1948). 'Boole and the Revival of Logic'. *Mind* LVII, pp. 149–75.

Knobloch, Eberhard (1987). 'Émile Borel as a Probabilist'. In Krüger, Gigerenzer and Morgan, eds. (1987), vol. 1, pp. 215–33.

Kolmogorov, Andrei Nicolaevich (1933). *Grundbegriffe der Wahrscheinlichkeitsrechnung*. Berlin: Springer. English edition *Foundations of the Theory of Probability*. New York: Chelsea, 1950.

Krüger, Lorenz; Gigerenzer, Gerd and Morgan, Mary S., eds. (1987). *The Probabilistic Revolution*. 2 volumes. Cambridge, Mass.: MIT Press.

Kuipers, Theo A. F. (2000). *From Instrumentalism to Constructive Empiricism*. Dordrecht-Boston: Kluwer.

Kyburg, Henry jr. (1968). 'The Rule of Detachment in Inductive Logic'. In Lakatos, ed. (1968), pp. 98–165.

Kyburg, Henry jr. and Smokler Howard, eds. (1964). *Studies in Inductive Logic and Probability*. New York - London - Sydney: Wiley. Second modified edition Huntington (N.Y.): Krieger, 1980.

Lakatos, Imre, ed. (1968). *The Problem of Inductive Logic*. Amsterdam: North-Holland.

van Lambalgen, Michiel (1987). *Random Sequences*. Academisch Proefschrift, University of Amsterdam.

Laplace, Pierre Simon (1774). 'Mémoire sur la probabilité des causes par les événements'. *Mémoires présentées à l'Académie des Sciences* VI, pp. 621–56.

Laplace, Pierre Simon (1814). *Essai philosophique sur les probabilités*. Paris: Courcier. English edition *Philosophical Essay on Probabilities*, edited and translated from the fifth French edition of 1825 by Andrew Dale. NewYork-Berlin-London: Springer, 1995.

Laplace, Pierre Simon (1878-1912). *Oeuvres complètes de Laplace*. 14 volumes. Paris: Gauthier-Villars.

Laudan, Larry (1973). 'Induction and Probability in the Nineteenth Century'. In *Logic, Methodology and Philosophy of Science IV*, ed. by Patrick Suppes, Leon Henkin, Athanase Joja, and Gr. C. Moisil. Amsterdam-London: North-Holland, pp. 429–38.

Leeds, Stephen (1984). 'Chance, Realism, Quantum Mechanics'. *The Journal of Philosophy* LXXXI, pp. 567–78.

Leibniz, Gottfried Wilhelm (1704). *Nouveaux essais sur l'entendement humain*. English edition *New Essays on Human Understanding*, translated and edited by Peter Remnant and Jonathan Bennett. Cambridge: Cambridge University Press, 1996.

Levi, Isaac (1967). *Gambling with Truth*. London: Routledge and Kegan Paul.

Levy, Paul (1979). *G.E. Moore and the Cambridge Apostles*. Oxford-New York: Oxford University Press.

Lewis, David (1980). 'A Subjectivist's Guide to Objective Chance'. In Jeffrey, ed. (1980), pp. 263–93.

Lindley, Dennis (1982). 'Bayesian Inference'. In *Encyclopedia of Statistical Sciences*, ed. by Samuel Kotz and Norman Johnson. New York: Wiley. Vol. 1, pp. 197–205.

Lindley, Dennis (1991). 'Sir Harold Jeffreys'. *Chance* IV, pp. 10–21.

Luckenbach, Sidney A., ed. (1972). *Probabilities, Problems, and Paradoxes*. Encino-

Belmont, California: Dickenson.

MacHale, Desmond (1985). *George Boole. His Life and Work*. Dublin: Boole Press.

Maistrov, Leonid Efimovich (1974). *Probability Theory. A Historical Sketch.* Translated from the Russian original (Moscow, 1967) by Samuel Kotz. New York-London: Academic Press.

Majer, Ulrich (1989). 'Ramsey's Conception of Theories: An Intuitionist Approach'. *History of Philosophy Quarterly* VI, pp. 233–58.

Majer, Ulrich (1991). 'Ramsey's Theory of Truth and the Truth of Theories: a Synthesis of Pragmatism and Intuitionism in Ramsey's Last Philosophy'. *Theoria* LVII, pp. 162–95.

Martin-Löf, Per (1969). 'The Literature on von Mises' Kollektivs Revisited'. *Theoria* XXXV, pp. 12–37.

Mayo, Deborah (1996). *Error and the Growth of Experimental Knowledge*. Chicago-London: The University of Chicago Press.

Mayr, Ernst (1982). *The Growth of Biological Thought*. Cambridge-London: Cambridge University Press.

Mazurkiewicz, Stefan (1932). 'Zur Axiomatik der Wahrscheinlichkeitsrechnung'. *Comptes rendues des séances de la Société des Sciences et des Lettres de Varsovie*, Classe III, XXV, pp. 1–4.

McCurdy, Christopher (1996). 'Humphreys's Paradox and the Interpretation of Inverse Conditional Propensities'. *Synthèse* CVIII, pp. 105–25.

McGuinness, Brian, ed. (1967). *Ludwig Wittgenstein und der Wiener Kreis*. Frankfurt am Main: Suhrkamp. English edition *Wittgenstein and the Vienna Circle. Conversations Recorded by Friedrich Waismann*. Oxford: Blackwell, 1979.

McGuinness, Brian (1982). 'Wittgenstein on Probability: A Contribution to Vienna Circle Discussions'. *Grazer Philosophische Studien* XVI-XVII, pp. 159–74.

McGuinness, Brian (1988). *Wittgenstein. A Life. Young Ludwig 1889-1921*. London: Duckworth. Penguin edition 1990.

Mellor, Hugh (1971). *The Matter of Chance*. Cambridge: Cambridge University Press.

Mellor, Hugh, ed. (1995). 'Better than the Stars'. *Philosophy* LXX, pp. 243–62.

Mill, John Stuart (1843). *A System of Logic Ratiocinative and Inductive*. London. New edition by John Robson in *Collected Works*. London: Routledge. Vol. 7, 1973; vol. 8, 1974. Reprinted 1996.

Miller, David (1994). *Critical Rationalism. A Restatement and Defence*. Chicago and La Salle, Ill.: Open Court.

Miller, David (1997). *Sir Karl Popper*. In *Biographical Memoirs of Fellows of the Royal Society* XLIII, pp. 367–409.

Milne, Peter (1986). 'Can there Be a Realist Single-Case Interpretation of Probability?'. *Erkenntnis* XXV, pp. 129–32.

Montmort, Pierre Rémond (1708). *Essay d'analyse sue les jeux de hasard*. Paris: Quilau. Second edition (including the correspondence) Paris, 1713.

Nagel, Ernest (1933). 'A Frequency Theory of Probability'. *The Journal of Philosophy* XXX, pp. 533–54.

Nagel, Ernest (1936a). 'The Meaning of Probability'. *Journal of the American Statistical Association* XXXI, pp. 10–26.

Nagel, Ernest (1936b). 'Critical Notices: *Wahrscheinlichkeitslehre*, by Hans Reichenbach'. *Mind* XLV, pp. 501–14.

Nagel, Ernest (1939a). *Principles of the Theory of Probability*. Chicago: The University of Chicago Press. Twelfth impression 1969.

Nagel, Ernest (1339b). 'Probability and the Theory of Knowledge'. *Philosophy of Science* VI, pp. 212–53.

Nagel, Ernest (1963). 'Carnap's Theory of Induction'. In Schilpp, ed. (1963), pp. 785–825.

Neyman, Jerzy (1977). 'Frequentist Probability and Frequentist Statistics'. *Synthèse* XXXVI, pp. 97–131.

Neyman, Jerzy and Pearson, Egon Sharpe (1967). *Joint Statistical Papers*. Cambridge: Cambridge University Press.

Niiniluoto, Ilkka (1998). 'Induction and Probability in the Lvov-Warsaw School'. In *The Lvov-Warsaw School and Contemporary Philosophy*. Dordrecht-Boston: Kluwer, pp. 323–36.

Olby, Robert (1981). "La Théorie génétique de la sélection naturelle vue par un historien". *Revue de Synthèse* CII, pp. 251–83.

Pascal, Blaise (1665). *Traité du triangle arithmétique*. Paris: Desprez. Reprinted in Pascal (1963), pp. 50–63.

Pascal, Blaise (1670). *Pensées*. Paris: Port Royal. English edition *Pascal's Pensées*, ed. by Martin Turrell. London: Harvill Press, 1962 (containing also the translation of the biographical sketch of Pascal written by his sister Gilberte Périer).

Pascal, Blaise (1963), *Oeuvres Complètes*, ed. by Louis Lafuma. Paris: Seuil.

Pearson, Egon (1938). *Karl Pearson: An Appreciation of Some Aspects of his Life and Work*. Cambridge: Cambridge University Press.

Pearson, Egon (1990). *'Student'. A Statistical Biography of William Sealy Gosset*. Oxford: Oxford University Press.

Pearson, Egon and Kendall, Maurice George, eds. (1970). *Studies in the History of Probability and Statistics*. London: Griffin.

Pearson, Karl (1892). *The Grammar of Science*. London: Walter Scott.

Pearson, Karl (1897). *The Chances of Death and Other Studies in Evolution*. 2 volumes. London: Edward Arnold.

Pearson, Karl, ed. (1914-30). *The Life, Letters and Labours of Francis Galton*. 3 volumes in 4 parts. Cambridge: Cambridge University Press.

Peirce, Charles Sanders (1910). 'Notes on the Doctrine of Chances'. In *Collected Papers of Charles Sanders Peirce*, ed. by Charles Hartshorne and Paul Weiss. Cambridge, Mass.: Harvard University Press, 1932, pp. 404–14.

Poincaré, Henri (1908). *Science et méthode*. Paris: Flammarion. English edition *Science and Method*. London: Thomas Nelson and Sons, 1914. Reprinted Bristol: Thoemmes, 1996.

Poincaré, Henri (1912a). *Calcul des probabilités*. Paris: Gauthier-Villars.

Poincaré, Henri (1912b). *Dernières pensées*. Paris: Flammarion. English edition *Mathematics and Science: Last Essays*. New York: Dover, 1963.

Poincaré Henri (1928-56). *Oeuvres de Jules Henri Poincaré*. 11 volumes. Paris: Gauthier-Villars.

Poisson, Siméon-Denis (1837). *Recherches sur la probabilité des jugements en matière criminelle et en matière civile. Précédées des règles générales du calcul des probabilités.* Paris: Bachelier.

Popper, Karl Raimund (1934). *Logik der Forschung.* Vienna: Springer. English enlarged edition *The Logic of Scientific Discovery.* London: Hutchinson, 1959, third revised edition 1968.

Popper, Karl Raimund (1938). 'A Set of Independent Axioms for Probability'. *Mind* XLVII, pp. 275–9. Reprinted with changes as 'Appendix *ii' in Popper (1934, English edition 1959, 1968), pp. 318–22.

Popper, Karl Raimund (1957). 'The Propensity Interpretation of the Calculus of Probability, and the Quantum Theory'. In *Observation and Interpretation*, ed. by Stephen Körner. London: Butterworths, pp. 65–70.

Popper, Karl Raimund (1959). 'The Propensity Interpretation of Probability'. *British Journal for the Philosophy of Science* X, pp. 25–42.

Popper, Karl Raimund (1962). *Conjectures and Refutations.* London: Routledge.

Popper, Karl Raimund (1974). 'Intellectual Autobiography'. In Schilpp, ed. (1974), pp. 3–181. Reprinted as *Unended Quest: an Intellectual Autobiography.* London: Fontana, 1976.

Popper, Karl Raimund (1982). *Quantum Theory and the Schism of Physics.* London: Hutchinson.

Popper, Karl Raimund (1983). *Realism and the Aim of Science.* London: Hutchinson.

Popper, Karl Raimund (1990). *A World of Propensities.* Bristol: Thoemmes.

Popper, Karl Raimund and Miller, David (1987). 'Why Probabilistic Support is not Inductive'. *Philosophical Transactions of the Royal Society of London*, A 321, pp. 569–91.

Porter, Theodore (1986). *The Rise of Statistical Thinking, 1820-1900.* Princeton: Princeton University Press.

Porter, Theodore (2004). *Karl Pearson. The Scientific Life in a Statistical Age.* Princeton: Princeton University Press.

Quetelet, Adolphe (1835). *Sur l'homme et le développement de ses facultés, ou Essai de physique sociale.* Paris: Bachelier. English edition *A Treatise on Man and the Development of his Faculties.* Edinburgh: Chambers, 1842.

Ramsey, Frank Plumpton (1922). 'Mr. Keynes on Probability'. *The Cambridge Magazine* XI, pp. 3–5. Reprinted in *The British Journal for the Philosophy of Science* XL (1989), pp. 219–22.

Ramsey, Frank Plumpton (1931). *The Foundations of Mathematics and Other Logical Essays*, ed. by Richard Bevan Braithwaite. London: Routledge and Kegan Paul.

Ramsey, Frank Plumpton (1990a). *Philosophical Papers*, ed. by Hugh Mellor. Cambridge: Cambridge University Press.

Ramsey, Frank Plumpton (1990b). 'Weight or the Value of Knowledge'. *British Journal for the Philosophy of Science* XLI, pp. 1–4 (with a 'Preamble' by Nils-Eric Sahlin).

Ramsey, Frank Plumpton (1991a). *Notes on Philosophy, Probability and Mathematics*, ed. by Maria Carla Galavotti. Naples: Bibliopolis.

Ramsey, Frank Plumpton (1991b). *On Truth*, ed. by Nicholas Rescher and Ulrich

Majer. Dordrecht-Boston: Kluwer.

Reichenbach, Hans (1925). 'Die Kausalstruktur der Welt und der Unterschied zwischen Vergangenheit und Zukunft'. *Sitzungsberichte, Bayerische Akademie der Wissenschaft*, Nov. 1925, pp. 133–75. English edition 'The Causal Structure of the World and the Difference between Past and Future', in Reichenbach (1978), vol. 2, pp. 81–119.

Reichenbach, Hans (1930). 'Kausalität und Wahrscheinlichkeit'. *Erkenntnis* I, pp. 158–88. English edition 'Causality and Probability'. In *Modern Philosophy of Science*, ed. by Maria Reichenbach, London: Routledge and Kegan Paul, 1959, pp. 67–78. Reprinted in Reichenbach (1978), vol. 2, pp. 333–44.

Reichenbach, Hans (1933). 'Die logischen Grundlagen des Wahrscheinlichkeitsbegriffs'. *Erkenntnis* III, pp. 401–25. English edition with modifications 'The Logical Foundations of the Concept of Probability'. In Feigl and Sellars, eds. (1949), pp. 305–23.

Reichenbach, Hans (1935). *Wahrscheinlichkeitslehre*. Leyden: Sijthoff. English expanded version *The Theory of Probability*. Berkeley-Los Angeles: University of California Press, 1949. Second edition 1971.

Reichenbach, Hans (1936). 'Logicist Empiricism in Germany and the Present State of its Problems'. *The Journal of Philosophy* VI, pp. 141–61.

Reichenbach, Hans (1937). 'La philosophie scientifique: une esquisse de ses traits principaux'. In *Travaux du IX Congrès International de Philosophie*, Paris: Hermann, pp. 86–91.

Reichenbach, Hans (1938). *Experience and Prediction*. Chicago-London: The University of Chicago Press.

Reichenbach, Hans (1956). *The Direction of Time*. Berkeley and Los-Angeles: University of California Press.

Reichenbach, Hans (1978). *Selected Writings, 1909-1953*, ed. by Maria Reichenbach and Robert S. Cohen. 2 volumes. Dordrecht-Boston: Reidel.

Reid, Constance (1982). *Neyman - from Life*. New York: Springer.

Roeper, Peter and Leblanc, Hugues (1999). *Probability Theory and Probability Logic*. Toronto-Buffalo-London: University of Toronto Press.

Runde, Jochen (1996). 'On Popper, Probabilities, and Propensities'. *Review of Social Economy* LIV, pp. 465–85.

Sahlin, Nils-Eric (1986). 'How to be 100% Certain 99.5% of the Time'. *The Journal of Philosophy* LXXXIII, pp. 91–111.

Sahlin, Nils-Eric (1990). *The Philosophy of F.P. Ramsey*. Cambridge: Cambridge University Press.

Sahlin, Nils-Eric and Rabinowicz, Wlodek (1998). 'The Evidentiary Value Model'. In *Handbook of Defeasible Reasoning and Uncertainty Management Systems*, ed. by Dov Gabbay and Philippe Smets. Dordrecht-Boston: Kluwer. Vol. 1, pp. 247–65.

Salmon, Wesley C. (1966). *The Foundations of Scientific Inference*. Pittsburgh: University of Pittsburgh Press. Reprinted 1967.

Salmon, Wesley C. (1967). 'Carnap's Inductive Logic'. *The Journal of Philosophy* LXIV, pp. 725–39.

Salmon, Wesley C. (1971). *Statistical Explanation and Statistical Relevance*.

Pittsburgh: University of Pittsburgh Press.

Salmon, Wesley C. (1979a). 'The Philosophy of Hans Reichenbach'. In *Hans Reichenbach: Logical Empiricist*, ed. by Wesley C. Salmon. Dordrecht-London: Reidel, pp. 1–84.

Salmon, Wesley C. (1979b). 'Propensities: A Discussion Review'. *Erkenntnis* XIV, pp. 183–216.

Salmon, Wesley C. (1981a). 'John Venn's *Logic of Chance*'. In *Probabilistic Thinking*, ed. by Jaakko Hintikka, David Gruender and Evandro Agazzi, Dordrecht: Reidel, pp. 125–38.

Salmon, Wesley C. (1981b). 'Robert Leslie Ellis and the Frequency Theory'. In *Probabilistic Thinking*, ed. by Jaakko Hintikka, David Gruender and Evandro Agazzi, Dordrecht: Reidel, pp. 139–43.

Salmon, Wesley C. (1984). *Scientific Explanation and the Causal Structure of the World*. Princeton: Princeton University Press.

Salmon, Wesley C. (1991). 'Hans Reichenbach's Vindication of Induction'. *Erkenntnis* XXXV, pp. 99–122.

Salmon, Wesley C. (1994). 'Carnap, Hempel and Reichenbach on Scientific Realism'. In *Logic, Language, and the Structure of Scientific Theories*, ed. by Wesley C. Salmon and Gereon Wolters. Pittsburgh-Konstanz: University of Pittsburgh Press-Universitätsverlag, pp. 237–54.

Salmon, Wesley C. (1998). *Causality and Explanation*. New York-Oxford: Oxford University Press.

Savage, Leonard Jimmie (1954). *The Foundations of Statistics*. New York: Wiley.

Schilpp, Paul Arthur, ed. (1963). *The Philosophy of Rudolf Carnap*. La Salle, Illinois: Open Court.

Schilpp, Paul Arthur, ed. (1974). *The Philosophy of Karl Popper*. La Salle, Illinois: Open Court.

Schneider, Ivo (1980). 'Why do we Find the Origin of a Calculus of Probabilities in the Seventeenth Century?'. In *Probabilistic Thinking*, ed. by Jaakko Hintikka, David Gruender and Evandro Agazzi, Dordrecht: Reidel, pp. 3–24.

Seidenfeld, Teddy (1979). *Philosophical Problems of Statistical Inference*. Dordrecht: Reidel.

Shea, William (2003). *Designing Experiments and Games of Chance. The Unconventional Science of Blaise Pascal*. Canton, Mass.: Science History Publications/USA.

Shimony, Abner (1992). 'On Carnap: Reflections of a Metaphysical Student'. *Synthèse* XCIII, pp. 261–74.

Shimony, Abner (1999). 'Philosophical and Experimental Perspectives on Quantum Physics'. In *Epistemological and Experimental Perspectives of Quantum Physics*, ed. by Daniel Greenberger, Wolfgang Reiter and Anton Zeilinger. Dordrecht-Boston: Kluwer, pp. 1–17.

Shimony, Abner (unpublished manuscript). 'The Family of Propensity Interpretations of Probability'. Text of a conference read at the Institute for Advanced Study in Bologna, 3/5/2003.

Skidelsky, Robert (1983-1992). *John Maynard Keynes*. 2 volumes. London:

Macmillan.

Skyrms, Brian (1966). *Choice and Chance*. Belmont, California: Dickenson. Fourth modified edition Belmont, California: Wadsworth, 2000.

Skyrms, Brian (1984). *Pragmatics and Empiricism*. New Haven-London: Yale University Press.

Skyrms, Brian (1996). 'The Structure of Radical Probabilism'. *Erkenntnis* XLV, pp. 286–97. Reprinted in Costantini and Galavotti, eds. (1997), pp. 145–57.

Skyrms, Brian (1990). *The Dynamics of Rational Deliberation*. Cambridge, Mass.: Harvard University Press.

Stadler, Friedrich (1997). *Studien zum Wiener Kreis. Ursprung, Entwicklung and Wirkung des Logischen Empirismus im Kontext*. Frankfurt am Main: Suhrkamp. English edition *The Vienna Circle. Studies in the Origins, Development, and Influence of Logical Empiricism*. Vienna-New York: Springer, 2001.

Stigler, Stephen (1986). *The History of Statistics. The Measurement of Uncertainty before 1900*. Cambridge, Mass.-London: Harvard University Press.

Stigler, Stephen (1999). *Statistics on the Table. The History of Statistical Concepts and Methods*. Cambridge, Mass.-London: Harvard University Press.

Suppes, Patrick (1970). *A Probabilistic Theory of Causality*. Amsterdam: North-Holland.

Suppes, Patrick (1981). 'The Limits of Rationality'. *Grazer philosophische Studien* XII-XIII, pp. 85–101.

Suppes, Patrick (1983). 'The Meaning of Probability Statements'. *Erkenntnis* XIX, pp. 397–403.

Suppes, Patrick (1984a). *Probabilistic Metaphysics*. Oxford: Blackwell.

Suppes, Patrick (1984b). 'Conflicting Intuitions about Causality'. In *Midwest Studies in Philosophy*, vol. 9, ed. by Peter A. French, Theodore E. Uehling and Howard K. Wettstein. Notre Dame, Indiana: University of Notre Dame Press, pp. 151–68. Reprinted in Suppes (1993a), pp. 121–37.

Suppes, Patrick (1991). 'Indeterminism or Instability: Does it Matter?'. In *Causality, Method and Modality*, ed. by Gordon G. Brittan jr. Dordrecht-Boston: Kluwer, pp. 5–22. Reprinted in Suppes (1993a), pp. 247–59.

Suppes, Patrick (1993a). *Models and Methods in the Philosophy of Science*. Dordrecht-Boston: Kluwer.

Suppes, Patrick (1993b). 'The Transcendental Character of Determinism'. In *Midwest Studies in Philosophy*, vol. 18, ed. by Peter A. French, Theodore E. Uehling and Howard K. Wettstein. Notre Dame, Indiana: University of Notre Dame Press, pp. 242–57.

Suppes, Patrick (1999). 'Nagel, Ernest'. In *American National Biography*, ed. by John A. Garraty and Mark C. Carnes. New York: Oxford University Press, pp. 216–8.

Suppes, Patrick (2002). *Representations and Invariance of Scientific Structures*. Stanford: CSLI Publications.

Suppes, Patrick and Zanotti, Mario (1996). *Foundations of Probability with Applications*. Cambridge-New York: Cambridge University Press.

Swinburne, Richard, ed. (2002). *Bayes's Theorem*. Oxford: Oxford University Press.

ter Hark, Michel (2002). 'Between Autobiography and Reality: Popper's Inductive

Years'. *Studies in History and Philosophy of Science* XXXIII, pp. 79–103.

ter Hark, Michel (2004). *Popper, Otto Selz and the Rise of Evolutionary Epistemology.* Cambridge: Cambridge University Press.

Todhunter, Isaac (1865). *A History of the Mathematical Theory of Probability from the Time of Pascal to that of Laplace.* Cambridge-London: Macmillan. Third edition New York: Chelsea, 1965.

Urbach, Peter (1987). *Francis Bacon's Philosophy of Science.* Chicago: Open Court.

van Brakel, Jan (1976). 'Some Remarks on the Prehistory of the Concept of Statistical Probability'. *Archive for History of Exact Sciences* XVI, pp. 119–36.

Venn, John (1866). *The Logic of Chance.* London-Cambridge: Macmillan. Second modified edition 1876, third modified edition 1888.

Venn, John (1870). *On Some of the Characteristics of Belief Scientific and Religious.* London-Cambridge: Macmillan.

Venn, John (1881). *Symbolic Logic.* London: Macmillan. Second modified edition 1884. Reprinted New York: Chelsea, 1971.

Venn, John (1889). *The Principles of Empirical or Inductive Logic.* London: Macmillan. Second modified edition 1907.

von Mises, Richard (1928). *Wahrscheinlichkeit, Statistik und Wahrheit.* Vienna: Springer. English edition *Probability, Statistics and Truth.* London-New York: Allen and Unwin, 1939. Reprinted New York: Dover, 1957.

von Mises, Richard (1939). *Kleines Lehrbuch der Positivismus.* The Hague. English edition *Positivism*, Harvard: Harvard University Press, 1951. Reprinted New York: Dover, 1968.

von Mises, Richard (1964). *Mathematical Theory of Probability and Statistics*, ed. by Hilda Geiringer. New York: Academic Press.

von Plato, Jan (1994). *Creating Modern Probability.* Cambridge-New York: Cambridge University Press.

von Wright, Georg Henrik (1982). *Wittgenstein.* Minneapolis: University of Minnesota Press.

Waismann, Friedrich (1930). 'Logische Analyse der Wahrscheinlichkeitsbegriffs'. *Erkenntnis* I, pp. 228–48. English edition 'A Logical Analysis of the Concept of Probability', in *Philosophical Papers*, ed. by Brian McGuinness. Dordrecht-Boston: Reidel, 1977, pp. 4–21.

Wald, Abraham (1937). 'Die Wiederspruchsfreiheit des Kollektivsbegriffes der Wahrscheinlichkeitsrechnung', *Ergebnisse eines mathematischen Kolloquiums* VIII, pp. 38–72.

Whewell, William (1840). *The Philosophy of the Inductive Sciences, Founded upon their History.* London: Parker. Reprinted London: Thoemmes, 1996.

Wittgenstein, Ludwig (1921). *Logisch-Philosophische Abhandlung. Annalen der Naturphilosophie.* First English edition (translated by Charles K. Ogden and Frank Plumpton Ramsey) *Tractatus Logico-Philosophicus.* London: Routledge and Kegan Paul, 1922. Second edition (translated by David F. Pears and Brian McGuinness) London: Routledge and Kegan Paul 1961, reprinted 2000.

Wrinch, Dorothy and Jeffreys, Harold (1919). 'On Some Aspects of the Theory of Probability', *Philosophical Magazine* XXXVIII, pp. 715–31.

Wrinch, Dorothy and Jeffreys, Harold (1921-23). 'On Certain Fundamental Principles of Scientific Inquiry'. *Philosophical Magazine, Part I*, XLII, pp. 369–90; *Part II*, XLV, pp. 368–74.

Wrinch, Dorothy and Jeffreys, Harold (1923). 'The Theory of Mensuration'. *Philosophical Magazine* XLVI, pp. 1–22.

Zabell, Sandy (1982). 'W.E. Johnson's "Sufficientness" Postulate'. *The Annals of Statistics* X, pp. 1091–9.

Zabell, Sandy (1989). 'The Rule of Succession'. *Erkenntnis* XXXI, pp. 283–321.

Zabell, Sandy (1991). 'Ramsey, Truth and Probability'. *Theoria* LVII, pp. 210–38.

Zabell, Sandy (1996). 'Confirming Universal Generalizations'. *Erkenntnis* XLV, pp. 267–83. Reprinted in Costantini and Galavotti, eds. (1977), pp. 127–44.

# *Index*

**A**

Aliotta, Antonio  210
Aristotle  125
Ayer, Alfred J.  195

**B**

Bacon, Francis  28-9, 34-5
Bar-Hillel, Yehoshua  152, 164
Barnard, George  17
Bayes, Thomas  2, 17-9, 27, 33, 45,
    47-51, 62-3, 96, 118-20, 143,
    179, 200, 209, 214-6, 222, 226
Berkeley, George  25, 146
Bernoulli, Daniel  16-7
Bernoulli, Jakob  13-5, 64
Bernoulli, Johann  15
Bernoulli, Nikolaus  15-6
Bertrand, Joseph  66-9
Boethius  125
Bohr, Niels  26
Boltzmann, Ludwig  26, 90, 160
Bolzano, Bernard  135-6, 141, 160
Bonhert, H.G.  164
Boole, George  3, 75, 77, 136-41,
    143, 147
Borel, Émile  15, 46, 191-4
Born, Max  26, 89
Bornet, Gérard  139
Braithwaite, Richard Bevan  145,
    195, 203

Broad, Charlie Dunbar  145, 154-5,
    158
Brown, Robert  25-6
Butts, Robert  36
Byrne, Edmund  7

**C**

Calderoni, Mario  210
Cameron, Laura  195
Campbell, Norman  180, 185, 204-6
Cantelli, Francesco  15, 110
Carabelli, Anna  152-3
Cardano, Gerolamo  9
Carnap, Rudolf  3, 45, 55, 63, 66,
    82, 92, 103, 135-6, 141, 147,
    150-2, 157-8, 162-78, 184, 199,
    203, 212, 216, 222-3, 225-7, 236
Chebyshev, Pafnuty Lvovich  15
Church, Alonzo  110, 129
Cicero  125
Clark, Peter  124
Cohen, Robert S.  164
Comte, Auguste  143
Condorcet, Marie-Jean-Antoine-
    Nicaolas Caritat, Marquis de
    18-9, 65, 133
Cook, Alan  179
Costantini, Domenico  4, 157
Cournot, Antoine Augustin  125
Cramér, Harald  46

## D

Dale, Andrew  18
d'Alembert, Jean  57, 65, 126
Darboux, Gaston  127
Darwin, Charles  21, 34
Daston, Lorraine  11-2, 16, 19, 33
David, Florence Nightingale  8-9,
    11, 15, 17
Dawid, Philip  53
de Broglie, Louis  26, 89
De Finetti, Bruno  4, 45, 49-50, 53,
    63, 66, 101, 112, 137, 158, 169,
    175, 177-8, 181-2, 194, 199-200,
    203, 208-226, 229, 231, 235-7
De Finetti, Fulvia  209
Democritus  125
de Moivre, Abraham  14-5, 17, 64
de Montmort, Pierre Rémond  15
De Morgan, Augustus  3, 75, 77-8,
    136-7, 142, 190
De Morgan, Sophia Elizabeth  137
Dewey, John  101
de Witt, Jean Paul  11
d'Holbach, Paul Henri Thiry  126
Diderot, Denis  126
Di Maio, Concetta  200
Dirac, Paul  26
Dokic, Jérôme  207
Donkin, William Fishburn  78, 138,
    142, 189-90, 228
Doob, Joseph L.  46, 110, 129
Dubislav, Walter  82
Dummett, Michael  136
Dunnington, Guy Waldo  17

## E

Earman, John  33, 39, 48, 134, 228,
    230
Edgeworth, Francis Ysidro  23, 67
Einstein, Albert  26, 107
Ellis, Robert Leslie  3, 71-6
Engel, Pascal  207
Epicurus  125
Euler, Leonhard  17

## F

Feigl, Herbert  99, 164
Feller, William  46, 54
Fermat, Pierre  7-8, 10
Festa, Roberto  4, 108, 173
Fetzer, James  114, 116-7, 121, 124
Fisher, Ronald Aylmer  21-5, 180,
    183, 235
Forrester, John  195
Francis, Henry Thomas  75
Frank, Philip  82
Franklin, James William  9
Fréchet, Maurice  46, 212
Freud, Sigmund  195
Fuller, Jean Overton  28

## G

Galavotti, Maria Carla 157, 178,
    184, 195, 200, 202-3, 208, 217,
    219, 225, 230, 237
Galilei, Galileo  9
Galton, Francis  21-3, 142
Garber, Daniel  9
Garbolino, Paolo  4
Gauss, Carl Friedrich  17
Geiringer, Hilda  82
Gibbs, Josiah Williard  26, 183, 204
Giere, Ronald  114-5, 124
Gigerenzer, Gerd  24, 26
Gillies, Donald  4, 61, 106-8, 110,
    114, 116-21, 123, 147-8, 153,
    235
Gillispie, Charles Coulston  13, 58
Gnedenko, Boris Vladimirovich  39,
    46
Good, Irving John  50, 158, 200,
    225, 229
Goodwin, Harvey  71
Gosset, William Sealy  23
Granger, Gilles-Gaston  19
Grattan-Guinness, Ivor  139
Graunt, John  11
Grelling, Kurt  82

## H

Hacking, Ian  8-9, 11-2, 20, 24-5, 99,

105, 114, 125, 135
Hacohen, Malachi  107
Hahn, Hans  57, 65, 82, 161, 164
Hailperin, Theodor  141
Hammond, Nicholas  8
Hampshire, Stuart  162
Harley, Robert  138
Harrod, Roy Forbes  144, 195
Hartshorne, Charles  164
Heidelberger, Michael  160, 162
Heisenberg, Werner  26-7, 89-90, 123, 231
Hempel, Carl Gustav  82, 89, 164
Herschel, John  34-6
Hilbert, David  155
Hintikka, Jaakko  152, 164, 173
Hobbes, Thomas  158
Holland, J.D.  10, 17
Hosiasson-Lindenbaum, Janina  166
Howie, David  24, 180-1, 195
Howson, Colin  50, 180, 228-30
Hull, David  34, 74
Hume, David  31-4, 37, 99, 146, 201, 214
Humphreys, Paul  118-9, 230
Huygens, Christiaan  10-1
Huzurbazar, Vasant S.  179

**I**

Ismael, Jenann  124

**J**

James, William  105, 202
Jaynes, Edwin T.  225
Jeffrey, Richard  4, 164-5, 174, 178, 190, 195, 210, 225-29, 233, 237
Jeffreys, Harold  3, 24, 45, 49, 150, 158, 166-7, 178-87, 195, 199, 211, 216, 222, 225, 236-7
Jevons, William Stanley  3, 136, 141-3
Johnson, Norman  50
Johnson, William Ernest  3, 63, 66, 145, 153-8, 169, 179, 183, 200, 212, 226
Jorland, Gérard  16

**K**

Kamlah, Andreas  160
Kant, Immanuel  2, 58
Kantola, Ilkka  7
Kaplan, David  164
Kendall, Maurice George  24, 130-1
Kenny, Anthony  159
Kepler, Johannes  30
Keynes, John Maynard  3, 45, 61, 101, 142, 144-55, 158, 167-8, 182-3, 191-5, 199, 200-1, 203-4, 222, 226, 228, 235
Keynes, John Neville  144
Keynes, Milo  145
Khinchin, Alexandr Yakovlevich  39, 46, 52
Kneale, William  139
Knobloch, Eberhard  192
Kolmogorov, Andrej Nikolaevich  2, 15, 46, 51-4, 110, 129
Kotz, Samuel  50
Krüger, Lorenz  26
Kuipers, Theo  108, 173
Kyburg, Hnery jr.  152

**L**

Laplace, Pierre Simon  2-3, 14, 17-20, 27, 34-5, 40, 50, 57-66, 68-9, 77, 110, 112, 126, 128, 132, 135, 141, 143, 148, 167, 184, 223, 226
Laudan, Larry  27, 143
La Vergata, Antonello  5
Leblanc, Hugues  55
Leeds, Stephen  124
Leibniz, Gottfried Wilhelm  8, 11-2, 135
Levi, Isaac  114
Levy, Paul  195
Lewis, David  229
Lindley, Dennis  50, 179, 181
Locke, John  2, 11, 146
Lyapunov, Alexandr Mikhailovich  15

**M**

Mach, Ernst  184, 210, 237

MacHale, Desmond   138
Maistrov, Leonid Efimovich   9, 13, 15, 17, 26, 52
Majer, Ulrich   195, 207
Markov, Andrej Andreevich   15, 46
Martin-Löf, Per   129
Maxwell, James Clerk   26
Mayo, Deborah   25
Mayr, Ernst   24
Mazurkiewicz, Stefan   54
McCurdy, Christopher   119
McGuinness, Brian   5, 159-61
McTaggart, John   144
Mellor, Hugh   114-5, 195
Mill, John Stuart   34-6, 74, 125-6, 143, 146
Miller, David   107, 114-7, 124
Milne, Peter   119
Moore, George Edward   144, 147, 195
Morgan, Mary   26
Morris, Charles   164

**N**

Nagel, Ernest   3, 66, 101-4, 230
Neurath, Otto   82, 101, 164
Newton, Isaac   30, 57
Neyman, Jerzy   25, 235, 238
Niiniluoto, Ilkka   166

**O**

Olby, Robert   24

**P**

Pacioli, Luca   9
Pascal, Blaise   7-10, 12, 81
Pearson, Egon   23-5, 238
Pearson, Karl   21-3, 25, 184
Peirce, Benjamin   105
Peirce, Charles Sanders   3, 102, 105-6, 133, 202, 207
Planck, Max   26, 89
Poincaré, Henri   3, 67, 126-9, 132-3, 191-2, 206
Poisson, Siméon Denis   8, 15, 19, 57, 64, 140-1
Popper, Karl Raimond   3, 45, 54, 102-3, 106-17, 123-4, 129-30, 133, 152, 235-6
Porter, Theodore   21, 23, 26, 105
Price, Richard   17, 27, 33, 49, 63, 143

**Q**

Quetelet, Adolphe   18, 20-1
Quine, Willard van Orman   164

**R**

Rabinowicz, Wlodek   14
Ramsey, Frank Plumpton   4, 101, 112, 144, 150, 152-3, 158, 167, 175, 177, 181-2, 184, 186-7, 190, 194-207, 209, 211-2, 217, 223, 226, 228-9, 235, 237
Ramsey, Lettice   195
Ramsey, Michael   195
Rees, Graham   28
Reichenbach, Hans   3, 48-9, 51, 87, 91-101, 103, 177, 222, 225, 233
Reichenbach, Maria   164
Reid, Constance   25
Reik, Theodor   195
Rescher, Nicholas   195
Richards, Ivor A.   195
Rilke, Rainer Maria   82
Roeper, Peter   55
Runde, Jochen   112
Russell, Bertrand   82, 92, 94, 107, 144-5, 159, 179, 195

**S**

Sahlin, Nils-Eric   5, 7, 14, 29, 195, 200, 207
Salmon, Merrilee   39
Salmon, Wesley   4, 31, 39, 48, 68, 73-4, 92, 99-101, 115, 118-9, 121, 152, 167, 170, 177
Savage, Leonard Jimmie   200, 210, 212, 215, 219, 229
Schilpp, Paul Arthur   103
Schlick, Moritz   82, 161, 164
Schneider, Ivo   9
Schnorr, Claus Peter   129
Schrödinger, Erwin   26, 89

Seidenfeld, Teddy 25
Shea, William 8
Shimony, Abner 122-4, 164, 175
Simpson, Thomas 17
Skidelsky, Robert 144
Skyrms, Brian 29-30, 34, 39, 46,
    200, 228-9
Spencer, Herbert 22
Stadler, Friedrich 82, 161
Stegmüller, Wolfgang 164
Stigler, Stephen 21, 23, 61-2, 64,
    67
Suppes, Patrick 4, 63-4, 68, 101,
    121-2, 124, 133-4, 152, 178,
    230-2, 237-8
Swinburne, Richard 50
Swirles, Bertha 179

**T**

Tartaglia, Niccoloò 9
ter Hark, Michel 107
Thomas Aquinas 7, 17, 27, 49, 75
Thomson, W. 78
Todhunter, Isaac 15, 19

**U**

Urbach, Peter 28, 50, 229-30

**V**

Vailati, Giovanni 210
van Brakel, Jan 9

van Lambalgen, Michiel 129
Venn, John 3, 35, 62, 73-81, 126,
    145, 183
von Kries, Johannes 84, 160, 162
von Mises, Ludwig 81
von Mises, Richard 3, 46, 53-4, 61,
    81-98, 101, 103, 117-8, 126, 128-
    30, 133, 166, 221-22, 235
von Neumann, Johann 237
von Plato, Jan 26, 46, 53, 129, 192
von Smoluchowski, Marian 26
von Wright, Georg 160

**W**

Waismann, Friedrich 3, 145, 158,
    160-4, 171
Wald, Abraham 110, 128-9, 221
Weldon, Raphael 21-2
Weyl, Hermann 155, 237
Whewell, William 34, 36
Whitehead, Alfred North 107
Wittgenstein, Ludwig 3, 145, 150,
    155, 158-62, 164, 195, 200-1,
    203
Wrinch, Dorothy 179-80, 183

**Z**

Zabell, Sandy 9, 63, 158, 173, 200
Zanotti, Mario 231, 232